SpringerBriefs in Water Science and Technology

For further volumes:
http://www.springer.com/series/11214

Yoshiaki Tsuzuki

Pollutant Discharge and Water Quality in Urbanisation

 Springer

Yoshiaki Tsuzuki
Engineering, Architecture and Information
   Technology (EAIT)
The University of Queensland
Brisbane, QLD
Australia

and

Research Centre for Coastal Lagoon
   Environments (ReCCLE)
Shimane University
Matsue
Japan

ISSN 2194-7244   ISSN 2194-7252   (electronic)
ISBN 978-3-319-04755-3   ISBN 978-3-319-04756-0   (eBook)
DOI 10.1007/978-3-319-04756-0
Springer Cham Heidelberg New York Dordrecht London

Library of Congress Control Number: 2014933666

Printed on acid-free paper

Springer is part of Springer Science+Business Media (www.springer.com)

Preface

In the last decade, there have been several attempts in developing a systematic approach in determination of appropriate interventions and/or measures for water quality control. One of the challenging approaches is to derive the pollution loading and its relationships to other socioeconomic factors in a given aquatic environment. This book encompasses an interesting initiative on the application of pollutant discharge per capita (PDC) as a key determinant of strategic development for water quality management and control. Not only theoretical considerations of PDC approach are given, the field experiences on PDC applications in several case studies in Japan and other developing countries are illustrated in the book chapters, which allow readers to realize the salient features of PDC as well as its advantages and limitations. The relationships of technological interventions and socioeconomic implications are clearly explained together with roles of potential stakeholders in implementing the recommended measures. Readers of the book will mainly benefit from a systematic understanding through integrative approach for water quality control.

Thammarat Koottatep

Acknowledgments

Most ambient water quality monitoring is conducted to evaluate water quality conformity with the ambient water quality standards. This is mostly for the administrative purpose. Chronological water quality alternations can be found from long-term water quality monitoring. Both horizontal and vertical water quality profiles in water bodies are important and interesting to understand the existence of pollutants and their distributions in the water environment. Water quality monitoring data can also be applied to estimate and evaluate pollutant load in the rivers. Land-based pollutants flowing into estuary and coastal sea can be evaluated using pollutant discharge parameters.

This book is based on several of my existing works and lecture notes. The lecture notes have been prepared for lectures in the Japan International Cooperation Agency (JICA) training course of water quality monitoring with participants from developing and middle-developed countries. My existing works include two chapter manuscripts and several scientific papers. One is on water quality profiles in urbanization (Tsuzuki 2010), another is on sanitation development and roles of Japan (Tsuzuki 2011). Scientific papers which contents are included in this book have been published in several journals including Science of the Total Environment, Journal of Environmental Sciences, Journal of Hydrology, Ecological Indicators, Water, Soil, and Air Pollution, Water Science and Technology, and Ecological Economics. The contents of the papers include basic concepts on municipal wastewater treatment, the Social Experiment Program in the Yamato-gawa River Basin, Japan, and relationship between economic development and pollutant discharge. The Social Experiment Program includes dissemination of soft measures in households to reduce pollutant discharge and to improve water environment. Quantitative evaluations have been conducted on several aspects including the effects of the soft and hard measures on river water quality improvement, and costs and benefits of the measures.

Contents of this book are (1) water quality in the rivers and coastal areas, (2) pollutant load and water quality, (3) soft measures in households, (4) relationship between economic development and pollutant discharge per capita (PDC), (5) municipal wastewater pollutant discharge control, and (6) water and sanitation in developing countries. During the processes and development of the research, some specific terminologies have been modified and different words have been applied for the same meanings. For example, "soft interventions" have been changed to "soft measures." "Environmental accounting housekeeping (EAH) books" have been changed to "pollutant discharge calculators." The research contents are mostly relevant to environmental engineering, while some research topics are relevant to river engineering, water quality sciences, or environmental economics. For the integrated river environment management, a comprehensive research in these fields is considered to be necessary. This book is useful to understand comprehensively a framework of river environment management especially the research on pollutant discharge and water quality related to urbanization.

The author has conducted lecture in the JICA training course from 2008 to 2012. The lecture note has been updated every year to include the latest research works. A book chapter has been published from NOVA Science Publisher in 2010 (Tsuzuki 2010), which has been prepared based on the JICA lecture note in 2008. A part of the contents of this book is based on the latest lecture note in 2012. The author has conducted the lecture by 2012 because the author has moved to Australia since 2013.

Anticipated readers are from university undergraduate and graduate students in the fields of environment engineering, environment economics, development economics, environment policy and related fields, and professionals and specialists in these fields. Some statistical data and information have been collected using the Internet. This will help readers to easily find further references in English and in Japanese. Some URL of the references may be different at the time of publication. Then, the readers can use authors' name and titles of references to search literatures and documents. For Japanese references, the readers should know authors' name and titles in Japanese when they try to find by using search engines. The readers can freely contact the author by e-mail in such cases. Moreover, a reference list in Japanese will be uploaded on the author's URL[1] after publication of the book.

These works have been conducted with supports to conduct research from many people especially when the author was in Japan as a Researcher at Shimane University in 2004–2007, and several positions at Toyo University in 2007–2009. First of all, I would like to thank Prof Tomonori Matsuo, Dr Takashi Mino, Prof Kiyoshi Toda and Assoc Prof Hisao Ohtake, the University of Tokyo (at that time), for supervising my study and research when I was a student in Japan, as well as the academics at The University of Queensland, Australia, especially Profs Hubert Chanson and David Lockington and Dr Badin Gibbes. I appreciate miscellaneous kinds of supports from many academics, researchers, students, administrative officers, government officers, industries, and environmental nongovernment organizations (NGOs) in Japan, Thailand, and Bangladesh, especially for Prof Hidenobu Kunii and Prof Yasushi Seike, Shimane University, Japan, Prof Minoru Yoneda, Kyoto University, Japan, Prof Hidetoshi Kitawaki and Prof Toshiya Aramaki, Toyo University, Japan, Prof Suraphong Wattanachira, Chiang Mai University, Thailand, Assoc Prof Thammarat Koottatep, Asian Institute of

---

[1] https://sites.google.com/site/yoshiakitsuzuki/.

Technology, Thailand, and Prof Md Mafizur Rahman, Bangladesh University of Engineering and Technology, Bangladesh. It has been possible to conduct a part of the research works based on several research funds especially which have been awarded to the author from two Japanese foundations, the Japan Education Centre of Environmental Sanitation, and Osaka Bay Wide Area Environment Centre and Solid Waste Research Foundation. I also thank the editorial staff of Springer and Scientific Publishing Services for helping this publication, including Miss Kiruthika Poomalai and Dhanusha M.

Some of the contents of this book including some figures and tables are reprinted or modified after obtaining copyright permission from the International Water Association (IWA), Japan Society of Civil Engineers (JSCE), Japan Society on Water Environment (JSWE), Japan Education Centre of Environmental Sanitation (JECES), Infrastructure Development Institute, Japan (IDIJ), NOVA Science Publishers, Elsevier, and Springer.

Brisbane, December 2013   Yoshiaki Tsuzuki

References

1. Tsuzuki Y (2010) Domestic wastewater pollutant discharge and pollutant load waterquality in the ambient water in developed and developing countries, (Chapter 5, 125–164). In: Kudret Ertuð and Ilker Mirza (eds) Water quality: physical, chemical and biological characteristics. Nova Science Publishers, Inc., New York, p 277. ISBN: 978-1-60741-633-3
2. Tsuzuki Y (2011) Sanitation development and roles of Japan, (Chapter 6, 179–202). In: Joel M. McManned (ed) Potable water and sanitation. Nova Science Publishers, Inc., New York, p. 266. ISBN: 978-1-61122-319-4

Contents

Abbreviations

BOD	Biological oxygen demand *or* Biochemical oxygen demand
CJ	Combined *johkasou* (*johkasou*) (*gappei-shori johkasou* in Japanese)
CO_2	Carbon dioxide
COD	Chemical oxygen demand
DO	Dissolved oxygen
E. coli	Escharchia coliform
EAH (book)	Environmental accounting housekeeping (book)
EKC	Environmental Kuznets curve
FC (F. coli)	Fecal (faecal) coliform
Habitat	United Nations Centre for Human Settlements, UNCHS
IDIJ	Infrastructure Development Institute of Japan
IWA	International Water Association
JECES	Japan Education Centre of Environmental Sanitation
JICA	Japan International Cooperation Agency
JSCE	Japan Society of Civil Engineers
JSWE	Japan Society on Water Environment
MDGs	Millennium Development Goals
MFA	Material flux (flow) analysis
MLIT	Ministry of Land, Infrastructure, Transport and Tourism of Japan
NGO	Non-government organization
NST (system)	Night soil treatment (system) (*kumitori-benjo* in Japanese)
PCD	Pollution Control Department, Thailand
PDC	Pollutant discharge per capita
PGC	Pollutant generation per capita
PLC_{wb}	Pollutant load per capita flowing into water body
POPs	Persistent organic pollutants
PPP-GNI	Purchase power parity based gross national income
PROSANEAR	National sewerage strategy project of Brazil funded by the World Bank
SJ	Simple *johkasou* (*tandoku-shori johkasou* in Japanese)
SSL	Seepage and septage of septic tank and leachate of composting
TC (T. coli)	Total coliform
TEPCO	Tokyo Electric Power Co. Ltd

TMDL	Total maximum daily load
TN	Total nitrogen
TP	Total phosphorus
UASB	Upflow anaerobic sludge blanket
UNCHS	United Nations Centre for Human Settlements (Habitat)
UNEP	United Nations Environment Programme
UNICEF	United Nations Children's Fund (former the United Nations International Children's Emergency Fund)
USA	United States of America
US-EPA	Environment Protection Agency of the USA
WHO	World Health Organization
WPCF	Water Pollution Control Federation
WWTPs	Waste water treatment plants

Chapter 1
Water Quality in the Rivers and Coastal Areas

Ambient water quality deteriorates with increases of population and industry activities in urban areas. One of the important urban and industrial water sources is surface water including rivers and lakes (Fig. 1.1). When we consider water intake from a river, water is utilized in the upper sections, for agriculture, industries and municipality usages. After these usages, wastewater is discharged after some treatment or without treatment into the river. The water is intaken and used again in the lower section. When pollutant discharge amounts are small enough, water can be easily used in the downstream. A Japanese old proverb says, "Water will be cleaner after flowing three *shaku* (about one meter)". However, when pollutant discharges excess certain amounts, the water cannot to be cleaned and causes water pollution. Ambient water is used for miscellaneous purposes include water supply, irrigation, navigation, fishery and recreation. Maintaining the water quality in ambient water is important to facilitate these utilisations.

Pollutants in municipal wastewater and ambient water are periodically or continuously monitored with water quality parameters including biological oxygen demand (BOD), chemical oxygen demand (COD_{Cr} or COD_{Mn}[1]), bacterial indicators (e.g. E. coli, F. coli and T. coli), nitrogen (e.g. NH_4–N, NO_2–N, NO_3–N, TK–N and TN), phosphorus (e.g. PO_4–P and TP), and other organic and inorganic pollutants including heavy metals (e.g. As, Cd, CN, Cr, Hg, Pb). There are water quality regulations controlling pollutant discharges and ambient water quality by countries and specific areas. Most commonly applied water quality parameters to regulate the ambient water quality in developed and developing countries are BOD, dissolved oxygen (DO) and bacterial indicators (e.g. total coliform, TC or T. coli, and fecal (faecal) coliform, FC or F. coli). Pollutant discharge control regulations of nutrients have also been enforced in some specific areas to control eutrophication with phytoplankton in fresh water, brackish water and sea water. Nutrient parameters and chlorophyll-a are monitored in these water bodies.

---

[1] In Japan, COD_{Mn} is usually used for administrative purposes.

Y. Tsuzuki, *Pollutant Discharge and Water Quality in Urbanisation*,
SpringerBriefs in Water Science and Technology,
DOI: 10.1007/978-3-319-04756-0_1, © The Author(s) 2014

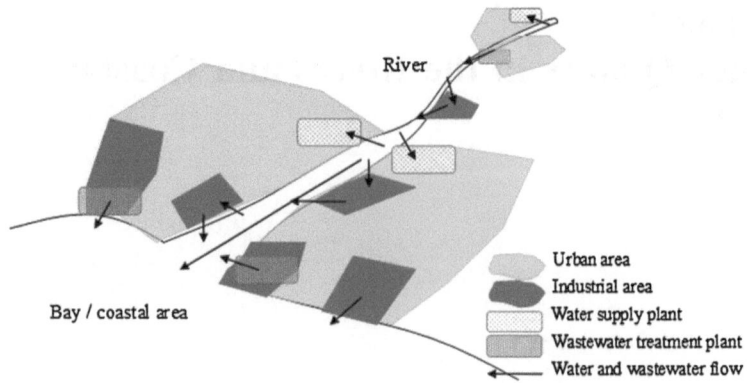

River

Bay / coastal area

Urban area
Industrial area
Water supply plant
Wastewater treatment plant
Water and wastewater flow

Fig. 1.1 Water and wastewater flow in the urban and industrial area along the river (Modified from Tsuzuki, 2010) (Copyright permission has been obtained from Nova Science Publishers)

1.1 Chronological Water Quality Changes

Chronological water quality changes are important to understand the alterations of water quality and to discuss the effects of natural and anthropogenic water pollutants. Some long-term chronological water quality alterations can be found in public and academic database and websites. For example, chronological water quality alterations in Lakes Shinji and Nakaumi, Japan, are found at the Environmental Database of the Lakes Shinji and Nakaumi.[2] Lakes Shinji and Nakaumi are the largest brakish estuary lagoon in Japan. COD_{Mn} has been almost constant in 1966–1985 with the alkali method, whereas COD_{Mn} decreased in 1983–1998 with the acid method.[3] Overall decreases of TP and TN have been observed at almost all the monitoring points in this area in 1976–1998.[4] These water quality trends did not directly reflect municipal wastewater pollutant discharge alterations. This is because there are other pollutant sources including other point sources, industries, and non-point sources, agriculture and forest, as well as inner production using nutrients in the water and sediments. Actually, the reason why water quality does not improve to the water quality levels in the 1960s, when we can swim in most of the ambient water, even after large pollutant discharge reductions using huge amounts of money is questionable and a target to be solved by water quality professionals.

Figures 1.2 and 1.3 show chronological water quality alterations at all of the official water quality monitoring points in Japan. Figure 1.2 shows alterations of conformity rate with the water quality standards in rivers, lakes and reservoirs, and coastal areas for organic carbons (BOD or COD_{Mn}) in 1974–2008. The conformity

---

[2] http://www.kisuiiki.shimane-u.ac.jp/envdb/waterquality.html (in Japanese and in English).

[3] http://www.kisuiiki.shimane-u.ac.jp/envdb/wq021.jpg

[4] http://www.kisuiiki.shimane-u.ac.jp/envdb/wq022.jpg

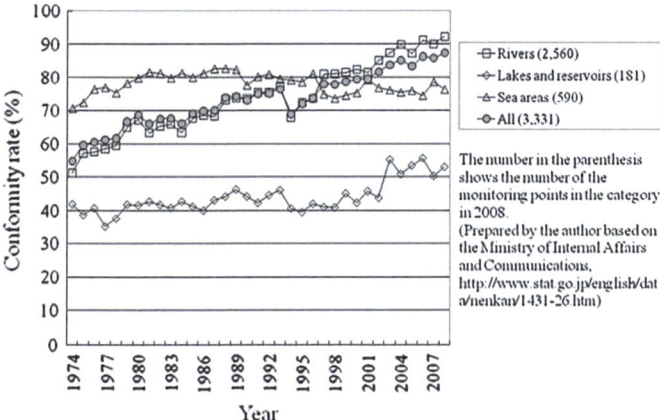

Fig. 1.2 Rates of conformity to water quality standards in 1974–2010 (Rivers and Lakes and reservoirs: BOD, and Sea areas: COD$_{Mn}$) (Prepared by the author based on Ministry of the Environment, Japan, 2013)

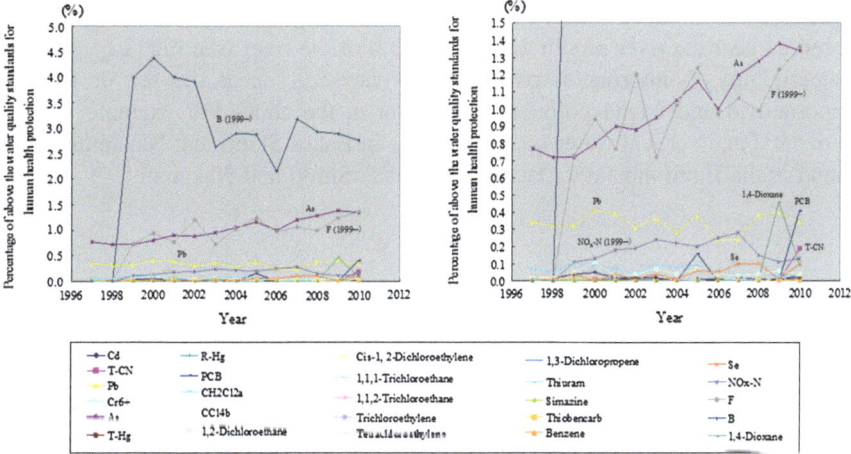

Fig. 1.3 Rates of non-conformity to water quality standards for protecting human health in 1997–2009 (Prepared by the author based on Ministry of the Environment, Japan, 2013)

rate of BOD in the rivers has improved from ca. 50 to ca. 90 %, that in lakes and reservoirs has improved from ca. 40 to ca. 50 %, and that of COD$_{Mn}$ in coastal areas has remained in the range of 70–80 %. Water quality regulations have been introduced around 1970 and water quality generally improved after 1970, during the period shown in the figure.

Figure 1.3 shows non-conformity rates to water quality standards of the parameters for protecting human health, which regulates toxic substances including heavy metals and persistent organic pollutants (POPs) in 1997–2009.

After the establishment and enforcement of the water quality regulations in the ambient water as well as pollutant discharge regulations of domestic and industrial wastewater, most toxic substances in the ambient water decreased to acceptable levels in a decade. Some toxic concentration increases have observed after 2000 including arsenic, lead, mercury, fluorine and boron. The reasons for these concentration increases have been considered as natural causes, the effects of sea water at the monitoring points in tidal areas, and unknown reason.

1.2 Horizontal and Vertical Water Quality Profiles

1.2.1 Horizontal and Vertical Water Quality Profiles in the Estuary in Japan

In order to estimate pollutant discharge contribution to pollutant load flowing into the water body, i.e. sea and lakes, the effect of eutrophication should be considered. Vertical water quality profiles in the river estuary, coastal areas and estuary lagoons are complicated, e.g. chlorophyll-a bloom or chlorophyll-a maximum is observed near the river mouth at certain depth of the river (see Fig. 1.5). Vertical water quality monitoring at river sections especially near the sea or lakes is important to understand pollutant movement in the areas. For example, several horizontal and vertical water quality profiles in Lakes Shinji and Nakaumi can be found at the Environmental Database of Lakes Shinji and Nakaumi.[5]

1.2.2 Horizontal and Vertical Water Quality Profiles in the Rivers in Developing Countries

Horizontal water quality is monitored in many countries. For example, the lower section of the Chao Phraya River is one of the water quality deteriorated rivers in Thailand (Fig. 1.4). Water quality deterioration is observed in Bangkok downtown area section of the Chao Phraya River for BOD, DO, nitrogen and phosphorus (Tsuzuki et al. 2008a, b[6]). Because of land flatness, salinity intrusion up to 50 km from the river mouth is observed depending on the rainy and dry seasons.

In the river section near the river mouth of the Tha Chin River (Fig. 1.4), salt water intrusion has been observed with vertical water quality measurement (Fig. 1.5). Chlorophyll-a maximum in 1–3 m depth near the river mouth has been observed. These horizontal and vertical water quality profiles should be considered

---

[5] http://www.kisuiiki.shimane-u.ac.jp/envdb/waterquality.html (in Japanese and in English).

[6] http://www.wsscc.org/resources/resource-publications/
water-quality-and-pollutant-load-ambient-water-and-domestic

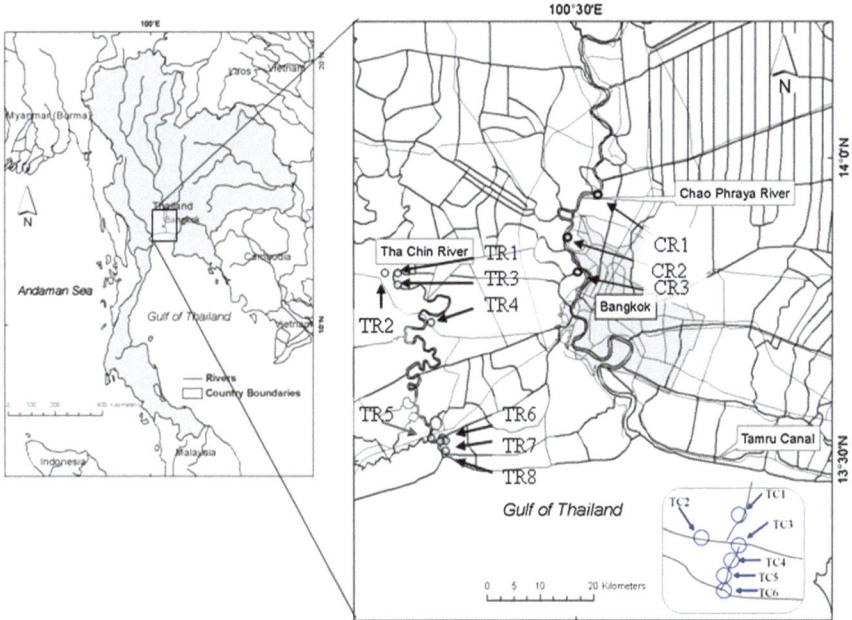

Fig. 1.4 Map around Bangkok, Thailand (Tsuzuki et al. 2008a, b) (Copyright permissions are obtained from JECES and JSCE)

in the pollutant load analysis, in which pollutant discharge effects on coastal water environment are analysed. Therefore, water quality monitoring plan and management for pollutant discharge control purposes should include observations of both horizontal and vertical water quality distributions especially deep water bodies and near the river mouths.

1.3 Groundwater Level and Subsidence

Groundwater level and subsidence are important under development conditions and much groundwater is used for water sources especially in urban areas. When field surveys have been conducted in Thailand and Bangladesh, the author has been impressed that ground water level decreases are still one of the major water environment problems in these countries. Subsidence has been experienced in Japan, however, the problem has been already mitigated by several methods including restrictions of groundwater uptake, transferring factories into sub-urban and rural areas from urban areas, and re-use of wastewater to decrease water usage amounts.

Water quality is determined by amounts of pollutant and water. When we consider water amounts, domestic and industrial water use amounts should be controlled in urban areas. In the industrial sector, large amounts of groundwater

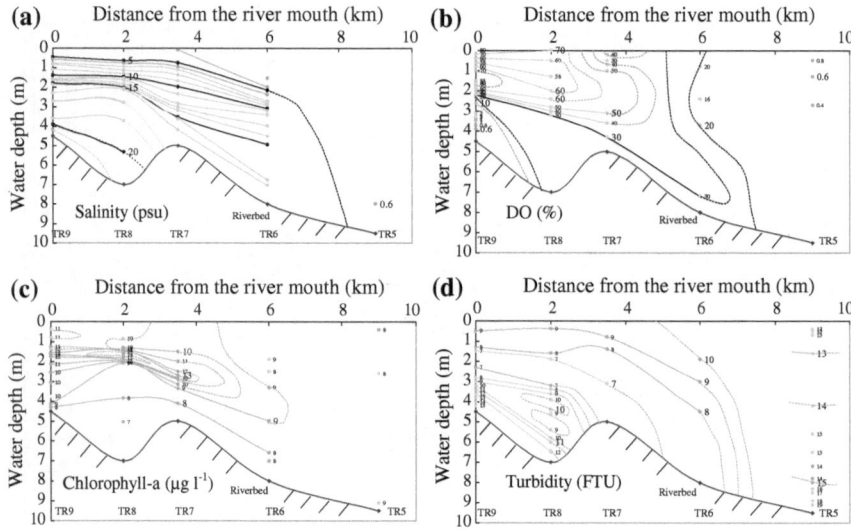

Fig. 1.5 Vertical water quality distribution near the river mouth of the Tha Chin River, observed on October 31, 2006 (Tsuzuki et al. 2009) (Copyright permissions has been obtained from JSWE)

have been used in urban areas in Japan, where many industries and factories have been built and developed. Such development caused subsidence especially in the 1960s. Some efforts on land use and industrial structure alterations have solved the subsidence problems in typical large urban areas including Tokyo, Nagoya and Osaka a few decades ago. Major industries and factories have been moved from urban areas to rural areas where industrial areas had been developed. Such solutions as groundwater withdrawal restrictions and moving industries from urban to rural areas have been conducted. Experiences of Japan in the 1960s will contribute to identification of the reasons and solutions of such problems in other countries.

In developing countries, groundwater level decreases of 1–3 m year^{-1} have been reported in urban areas, which should cause subsidence in the near future. Interventions including wastewater reuse, water supply source conversion to surface water, and land use planning especially for industries should be considered and implemented to alleviate subsidence.

References

Ministry of the Environment, Japan (2013) Water quality monitoring results of public water areas. http://www.env.go.jp/water/suiiki/index.html (in Japanese)

Tsuzuki Y, Koottatep T, Rahman MM, Ahmed F (2008a) Water quality in the ambient water and domestic wastewater pollutant discharges in the developing countries: possibility of combined johkasou exports, Joukasou Kenkyu. J Domest Wastewater Treat Res 20(1):1–13. (in Japanese with English abstract)

Tsuzuki Y, Koottatep T, Ahmed F, Rahman MM (2008b) Domestic wastewater pollutant discharge and pollutant load in the tidal area of the ambient water in developing countries: survey results in autumn and winter in 2006. J Glob Environ Eng 13:121–133. (Japan Society of Civil Engineers). http://www.wsscc.org/resources/resource-publications/water-quality-and-pollutant-load-ambient-water-and-domestic

Tsuzuki Y, Koottatep T, Rahman MM (2009) Water quality profiles of the tidal rivers and canal in per-urban of Bangkok, Thailand, and Dhaka, Bangladesh, focusing on the water quality transition in coastal areas, J Japan Soc Water Environ 32(1):47–52. (in Japanese with English abstract)

Tsuzuki Y (2010) Chapter 5: Domestic wastewater pollutant discharge and pollutant load water quality in the ambient water in developed and developing countries, 125–164, in Kudret Ertuð and Ilker Mirza eds. Water Quality: Physical, chemical and biological characteristics, Nova Science Publishers, Inc., p 277, ISBN: 978-1-60741-633-3

Chapter 2
Pollutant Load and Water Quality

Pollutant load per capita flowing into water body (PLC_{wb}) is an index to evaluate contribution of municipal wastewater pollutant discharge in pollutant load flowing into ambient water body such as coastal zones, bays and lakes. PLC_{wb} is not commonly applied in water pollution fields, however, PLC_{wb} should be more widely applied because of its easily understandable characteristics especially for comparison purposes among wastewater treatment methods or systems, among river sub-basins, and among countries, for miscellaneous stakeholders including policy makers, professionals of water environment and wastewater treatment and ordinary citizens.

PLC_{wb} is different by wastewater treatment methods and sub-river basins. For example, PLC_{wb}-BOD has been estimated to be 0.83 g person^{-1} day^{-1} (gpd) for WWTPs populations, 0.8–2.4 gpd for combined *johkasou* populations, 8.3–24 gpd for simple *johkasou* populations, and 7.8–21 gpd for night soil treatment system populations in a river basin of Japan (Tsuzuki 2006). The pollutant discharge calculator should be prepared for each sub-river basin and for each municipal wastewater treatment method. The pollutant discharge calculators can be utilised to enhance community involvement in water environment preservation and improvement.

Pollutant discharge per capita (PDC) is an index to simply indicate pollutant discharge amount to ambient water. PDC of municipal wastewater is a function of pollutant generation per capita (PGC) and treatment efficiencies of municipal wastewater treatment systems. The Millennium Development Goals (MDGs) sanitation indicator is 100 % in Japan. However, there have still been some municipal wastewater pollutant discharge problems. One of the key indices in sanitation and wastewater treatment in Japan is population with flush toilets. On the contrary, in Thailand, the MDGs sanitation indicator is 99–100 %, however, PDCs in Thailand are larger than those in Japan (Tsuzuki et al. 2009c). Therefore, more improvement of municipal wastewater treatment is necessary to improve the ambient water quality in Thailand. These situations may be similar in other developing countries with large MDGs sanitation indicator or on truck country.

In this chapter, estimation methods of PLC_{wb} and their results are explained. Water quality profiles in the past and present are explained to consider the relationships

Y. Tsuzuki, *Pollutant Discharge and Water Quality in Urbanisation*,
SpringerBriefs in Water Science and Technology,
DOI: 10.1007/978-3-319-04756-0_2, © The Author(s) 2014

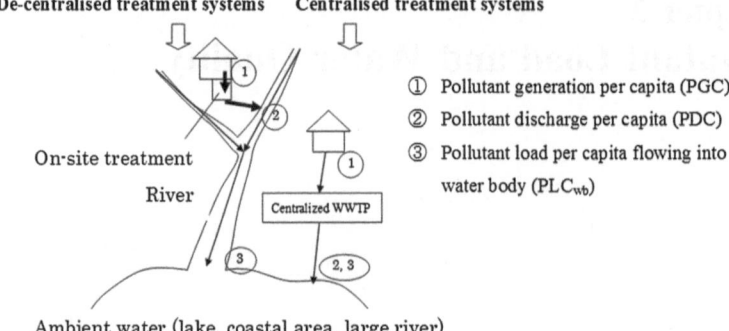

De-centralised treatment systems   Centralised treatment systems

① Pollutant generation per capita (PGC)

② Pollutant discharge per capita (PDC)

③ Pollutant load per capita flowing into
water body (PLC$_{wb}$)

On-site treatment

River

Centralized WWTP

Ambient water (lake, coastal area, large river)

Fig. 2.1 Water pollutant indicators in centralized and on-site wastewater treatment systems (Modified from Tsuzuki 2005). (Copyright permissions have been obtained from JECES and Nova Science Publishers)

between pollutant discharges and the ambient water quality. The sub-river basin basis analysis of PLC$_{wb}$ has been firstly conducted in the Miyako-gawa River Basin, Chiba Prefecture, Japan, which flows into the Tokyo Bay (Tsuzuki 2006). The indicator, PLC$_{wb}$, has been quantitatively evaluated by sub-river basin and wastewater treatment method. An example of pollutant discharge calculator of municipal wastewater shows the quantification concept and method of the soft intervention effects on pollutant discharge reductions and ambient water quality improvement.

2.1 Pollutant Load per Capita Flowing into Water Body (PLC$_{wb}$)

In this section, the municipal wastewater pollutant indicators and their relationships with ambient water are discussed and explained. There are several indicators on water pollutant discharges and environmental water quality (Fig. 2.1):

1. Pollutant generation per capita (PGC);
2. Pollutant discharge per capita (PDC);
3. Pollutant load per capita flowing into water body (PLC$_{wb}$); and
4. Water quality.

The water body is considered as the ambient water especially bay or coastal areas in Fig. 2.1. Analysis of PLC$_{wb}$ and Pollutant discharge calculator of municipal wastewater is conducted for several on-site and centralised municipal wastewater treatment systems in Japan (Fig. 2.2).

PDC and PLC$_{wb}$ are calculated with Eqs. 2.1 and 2.2

$$(PDC(POL_j))_i = PGC_j \times \frac{(100 - WTE_{i,j})}{100} \qquad (2.1)$$

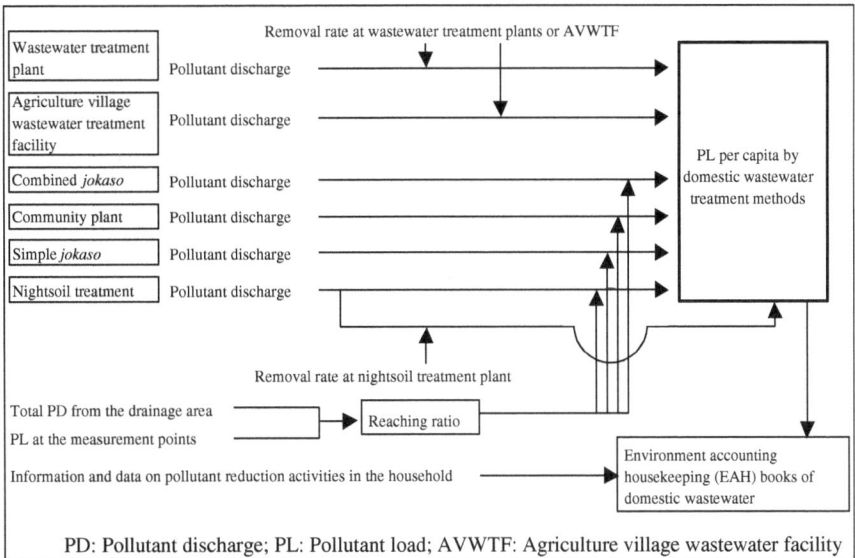

PD: Pollutant discharge; PL: Pollutant load; AVWTF: Agriculture village wastewater facility

Fig. 2.2 Flow chart of the analysis of pollutant load per capita flowing into the water body (PLC$_{wb}$) and pollutant discharge calculator for municipal wastewater. (Tsuzuki 2006) (Removal rate is pollutant removal efficiency. Reaching ratio is pollutant yield rate). (Copyright permissions have been obtained from Elsevier and Nova Science Publishers)

where $(PDC(POL_j))_i$ is pollutant discharge per capita (PDC) of pollutant j with municipal wastewater treatment method i (g person^{-1} day^{-1}); PGC_j is pollutant generation per capita (PGC) of pollutant j (g person^{-1} day^{-1}); and $WTE_{i,j}$ is wastewater treatment efficiency of pollutant j with municipal wastewater treatment method i (%).

$$(PLC_{wb}(POL_j))_i = (PDC(POL_j))_i \times \frac{PRY_j}{100} \tag{2.2}$$

where $(PLC_{wb}(POL_j))_i$ is pollutant load per capita flowing into water body (PLC$_{wb}$) of pollutant j with municipal wastewater treatment method i (g person^{-1} day^{-1}); and PRY_j is pollutant runoff yield of the pollutant j (%).

Pollutant discharge within a sub-river basin, $PD_{ka,j}$ (g day^{-1}), is calculated for the area above the monitoring point using Eq. 2.3 (Tsuzuki 2006).

$$PD_{ka,j} = PD_{k,j} \times \frac{A_{ka}}{A_k} \tag{2.3}$$

where $PD_{k,j}$ is pollutant discharge of pollutant j in the sub-river basin k (g day^{-1}), A_{ka} is land area above the monitoring point in the sub-river basin k (km^2) and A_k is total land area in the sub-river basin k (km^2).

For sub-river basins downstream of the other sub-river basin(s), pollutant load used for the calculation of pollutant runoff yield is calculated with Eq. 2.4, and corresponds to pollutant discharge.

$$PL_{k,j} = PL_{k-1,j} + PD_{k,j} \times \frac{A_{ka}}{A_k} \qquad (2.4)$$

where, $PL_{k,j}$ is pollutant load of pollutant j in the rivers at the monitoring points in the sub-river section k (g day^{-1}).

For river basins with a riverside purification facility, the pollutant discharge of the river basin has been calculated using the removal efficiency of a pollutant in the river water purification facility, and the treated and untreated river water volumes in the river water purification facility (Eq. 2.5).

$$PL_{kb,j} = PL_{k,j} \times \frac{\{(1 - \frac{RE_v}{100}) \times Q_t + Q_u\}}{Q_t + Q_u} \qquad (2.5)$$

where, $PD_{kb,j}$ is pollutant load of pollutant j below the river water purification facility v in the sub-river section k (g day^{-1}), $PD_{k,j}$ is pollutant load of pollutant j above the river water purification facility v in the sub-river section k (g day^{-1}), RE_v is removal efficiency of the riverside purification facility v (%), Q_t is treated river water volume in the river water purification facility (m^3 day^{-1}), and Q_u is untreated river water volume in the river water purification facility (m^3 day^{-1}).

In Eq. 2.5, it is assumed that only one river water purification facility is developed in the corresponding sub-river section. When there are more than one river water purification facilities in the sub-river section k, there are two calculation methods. One is calculations using Eq. 2.5 for the number of times of the facilities. Another is one-time calculation supposing the effects of several river water purification facilities into one specific removal efficiency with pollutant loads at the most upper and the lowest points of these river water purification facilities.

Figure 2.3 shows some places in Japan explained in this book. Tokyo Bay is surrounded by Chiba and Kanagawa Prefectures and Tokyo Metropolitan.

We can estimate contributions of municipal wastewater discharges to ambient water quality (Table 2.1). PLC$_{wb}$(BOD) has been found to be different by sub-river basin even with the same municipal wastewater treatment method in a case study in Chiba City, Japan (Fig. 2.4) (Tsuzuki 2006). Contribution of pollutant discharge to the Tokyo Bay can be estimated for each sub-river basin and municipal wastewater treatment method. Then, effects of pollutant discharge reduction measures or soft measures in households can be estimated for each sub-river basin and municipal wastewater treatment method (Table 2.1). PLC$_{wb}$ depended on sub-river basin even with the same wastewater treatment method.

Along with economic development, PGC should increase to some extent with increase of foods, chemicals and materials amounts used in people's lives and industries. PDC should increase with PGC increase when appropriate wastewater treatment systems are not applied, and decrease after wastewater treatment system

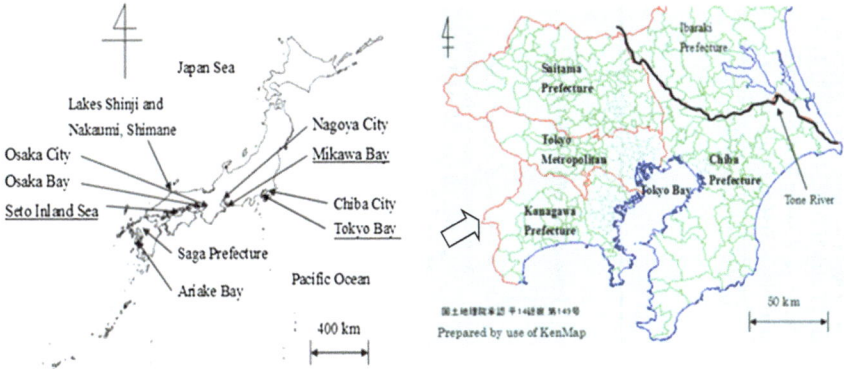

Fig. 2.3 Some specific places described in this book in Japan and around Tokyo Bay. There is the Tone River between Ibaraki and Chiba Prefectures, and most pollutants flowing into the Tokyo Bay are from Saitama, Chiba and Kanagawa Prefectures and Tokyo Metropolitan

development. PGC and PDC can be decreased with soft interventions. PLC_{wb} reflects PGC, PDC, pollutant removal efficiency of the wastewater treatment systems, and natural purification effect in ambient water. Water quality, which is commonly applied as the water environment indicator, is a function of pollutant load and water amount or flow rate in ambient water. Monitored water quality would be a base dataset to estimate PGC, PDC and PLC_{wb}. Water quality itself is important to make judgment for the appropriateness of the water usage for many purposes including water supply sources, irrigation, navigation, fishery, recreation, swimming and hydro power generation.

2.2   Roles of Fishery for Water Environment

How are pollutants discharged to coastal area decreased? Natural purification, discharging to neighbouring water body and removal by fishery are major possible reasons for pollutants reduction in coastal areas. Removal amounts of organic carbon (COD_{Mn} in this case), total nitrogen (TN) and total phosphorus (TP) by fishery have been estimated in the Tokyo Bay for the three periods of 1935, 1960, and 1997–2001 as shown in Fig. 2.5 (Tsuzuki 2004). Annual fishery amounts in the Tokyo Bay has been largest in 1960 among the three periods and decreased after industrialization and water pollution in the sea area. Total pollutant discharge amounts have been regulated in major enclosed sea areas in Japan, i.e. the Tokyo Bay, Mikawa Bay and Seto Inland Sea (Fig. 2.3). The Mikawa Bay is adjacent to Nagoya City and Seto Inland Sea includes the Osaka Bay. The concept of total pollutant discharge control in Japan is similar to total maximum daily load (TMDL) regulations in the USA. However, the number of subjected areas in Japan is limited. In the fifth total pollutant discharge reduction plan in Tokyo Bay area in

Table 2.1 An example of environmental accounting housekeeping (EAH) books or pollutant discharge calculators for municipal wastewater: population served with simple johkasou (SJ) and the river basin of the Miyakogawa River, Chiba City, Japan (Tsuzuki 2006) (see Fig. 2.4). The pollutant discharge calculators are prepared for wastewater treatment methods and sub river basins. Pollutant load decrease flowing into Tokyo Bay is estimated by applying the pollutant discharge calculators. (Copyright permissions have been obtained from Elsevier and Nova Science Publishers)

Miyakogawa River, upper drainage area Simple *jokaso*	Pollutant load ratio[a]				Pollutant load flowing into public water body and pollutant reduction effects[b]				Daily life				A month, 30 days				Estimation for calcuation
	BOD %	COD %	TN %	TP %	BOD mg	COD mg	TN mg	TP mg	BOD mg	COD mg	TN mg	TP mg	BOD g	COD g	TN g	TP g	
Nightsoil[c]																	
Bath[c]	(20)	(15)	(20)	(60)	840	3120	2640	300									
Decrease shampoo and soap[d]					1530	1740	130	80	460	520	39	24	14	16	1.2	0.7	
					460	520	39	24	460	520	39	24	14	16	1.2	0.7	The decrease effect to be 30 %
Kitchen[c]	(60)	(70)	(50)	(30)	4590	8110	330	40	320	570	70	2	10	17	2.1	0.1	The previous used amount to be 5 ml person^{-1} day^{-1} (2 g-BOD, 2 g-COD, 80 mg-TN and 0 g-TP person^{-1} day^{-1})
No use of detergent[d]					530	1780	35	0									Decrease to half
Decrease detergent[d]					265	890	18	0									Pollutant loads of rice washing water to be 2 g-BOD, 2 g-COD, 24 mg-TN and 2 mg-TP person^{-1} day^{-1}
Do not drain rice washing water[d]					530	1780	11	1									
Use paper filter for kitchen[d]					320	570	70	2	320	570	70	2	10	17	2.1	0.1	Removal rate: BOD:7 %, COD:7 %, T-N:21 %, and TP: 4 %
Use net for kitchen[d]					140	240	50	1									Removal rate: BOD:3 %, COD:3 %, T-N:15 %, and TP: 2 %
Treatment during and after cooking[d]					2300	4060	170	20									Removal rate: 50 %
Do not drain residual liquid Dressing 5ml[d]					2940	n.a.[e]	n.a.	n.a.									BOD: 660,000 mg l^{-1}, and waste amount: 5 ml

(continued)

Table 2.1 (continued)

Miyakogawa River, upper drainage area Simple *jokaso*	Pollutant load ratio[a]				Pollutant load flowing into public water body and pollutant reduction effects[b]				Daily life				A month, 30 days				Estimation for calcualtion
	BOD %	COD %	TN %	TP %	BOD mg	COD mg	TN mg	TP mg	BOD mg	COD mg	TN mg	TP mg	BOD g	COD g	TN g	TP g	
Chinese noodle soup 50ml[d]					1160	n.a.	n.a.	n.a.									BOD: 26,000 mg l^{-1}, and waste amount: 50 ml
Used edible oil 10ml[d]					14900	n.a.	n.a.	n.a.									BOD: 1,670,000 mg l^{-1}, and waste amount: 10 ml
Washing clothes[c]	(20)	(15)	(30)	(10)	1530	1740	200	10									
Decrease detergent to half[d]					690	780	90	5									The contributions of detergent to be 90 % of pollutant loads by washing clothes
Total pollutant loads[f]	(100)	(100)	(100)	(100)	8500	14700	3300	430	7720	13610	3191	404	232	408	95.7	12.1	
Decrese of pollutant load per capita									780	1090	109	26	23	33	3.3	0.8	
Decrese of pollutant load for a family of four													94	131	13.1	3.1	

[a] Pollutant load ratios without nightsoil, determined by the author based on the data in Chiba Prefectural Institute for Water Quality Preservation (1980)

[b] Pollutant load by households activities calculated at the measurement point nearest the river mouth, and corresponding pollutant reduction effects of measures in the households

[c] Pollutant load flowing into the water body by pollutant load discharge categories in the households

[d] Pollutant load reduction measures in the households and their effect estimations calculated by use of pollutant load per capita flowing into the water body or reaching ratio

[e] Not available because of data or information dificiency

[f] Total pollutant load per capita flowing into the wate⁻ body, PLC-FW

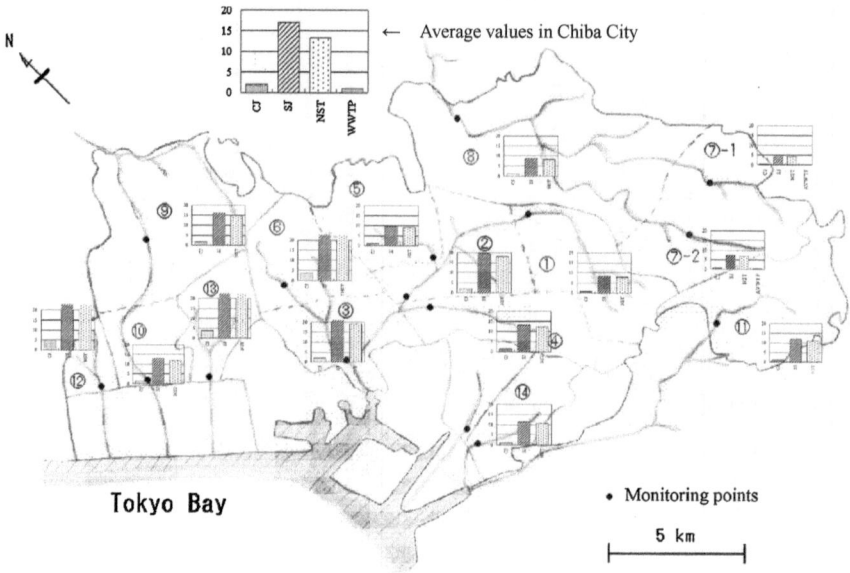

Fig. 2.4 PLC$_{wb}$(BOD) by sub-river basins and wastewater treatment methods in Chiba City, Japan (g-BOD person^{-1} day^{-1}) (Tsuzuki 2006). (Copyright permissions have been obtained from Elsevier and Nova Science Publishers)

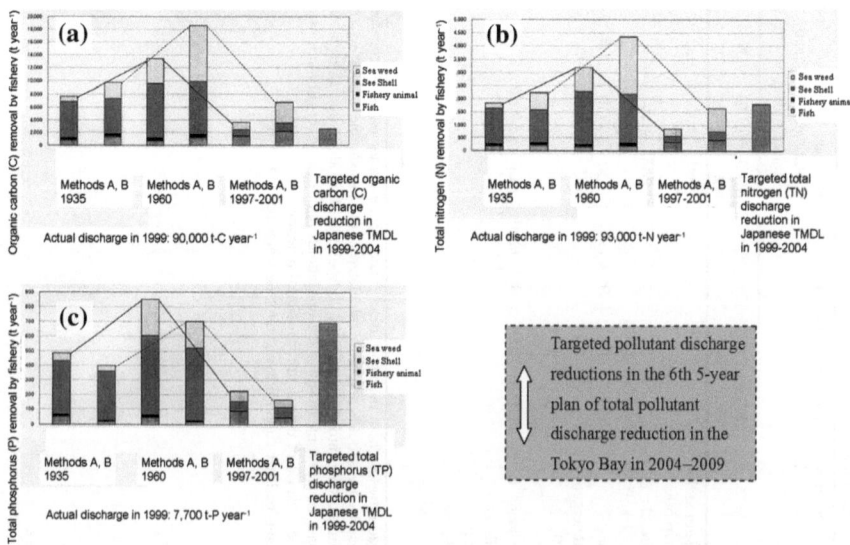

Fig. 2.5 Pollutant removal amounts by fishery in Tokyo Bay in 1935, 1960 and 1997–2001 **a** Organic carbon (C) **b** total nitrogen (TN), and **c** total phosphorus (TP) (Modified from Tsuzuki 2004)

1999–2004, targeted pollutant discharge reduction amounts have been 19 t-COD_{Mn} day^{-1}, 5 t-TN day^{-1}, and 1.9 t-TP day^{-1}, or 6,900 t-COD_{Mn} year^{-1}, 1,800 t-TN year^{-1} and 690 t-TP year^{-1} (Table 2.2). Targeted reduction amounts of COD_{Mn}, TN and TP discharged into the Tokyo Bay in 2004–2009 in the 6th Plan have been within the same order as those in the 5th Plan. The actual discharged amounts of COD_{Mn}, TN and TP have been less than the targeted amounts in the 5th Plan, and the Plan has been considered to be achieved in a quite good extents. For example, COD_{Mn} discharge amount has been targeted to be 228 t day^{-1} in 2004 in the 5th Plan schemes, and the actual discharged amount has been 211 t day^{-1} in 2004 (Table 2.2).

These targeted amounts are almost the same range of yearly organic carbon, TN and TP removal amounts by fishery in 1960 (Fig. 2.5). Pollutant removal by fishery is not neglectable amount in regards to pollutant discharge and removal in the Tokyo Bay area.

Table 2.2 Pollutant discharge regulated amounts in the 5th and 6th total pollutant discharge reduction plan in Tokyo Bay area in 1999–2004 and in 2004–2009.

2.3 Relationship Between Pollutant Discharge and Water Quality: Perturbation or Hysteresis

When the amounts of pollutants discharged to the ambient water including the rivers, lakes and coastal areas, water quality of the ambient water is generally considered to deteriorate and vice versa. The relationship between pollutant discharge amounts and ambient water quality is generally considered to be something like linear or first-order relationship. Miscellaneous measures are conducted to reduce pollutant discharge and to improve ambient water quality.

On the contrary, BOD in the Yamato-gawa River, Japan, has been found to deteriorate more than expected when pollutant discharge has rapidly increased in the late 1960s and early 1970s (Figs. 2.6 and 2.7) (Tsuzuki 2013). In phase I (Fig. 2.7d), original relationship between pollutant discharge and water quality has been considered to be a linear, however, water quality has deteriorated because of rapid increase of pollutant discharge over a threshold level in the river basin (Phase II). After several years, water quality has improved besides continuous increase of pollutant discharge, water quality has improved and the relationship has become close to the original linear relationship with pollutant discharge decrease (Phase IV), and the relationship has returned to the original linear relationship (Phase V). It has taken long time when water quality has once deteriorated cause by rapid pollutant discharge increase or pollutant discharge excess a threshold level. The relationship is similar to perturbation or dynamic equilibrium change of stable conditions in the fields of ecosystems.

Table 2.2 Pollutant discharge regulated amounts in the 5th and 6th total populant discharge reduction plan in Tokyo Bay area in 1999–2004 and in 2004–2009

Local government	Unit	COD_Mn				TN				TP			
		5th Plan		6th Plan		5th Plan		6th Plan		5th Plan		6th Plan	
		Discharge in 1999	Target in 2004	Discharge in 2004	Target in 2009	Discharge in 1999	Target in 2004	Discharge in 2004	Target in 2009	Discharge in 1999	Target in 2004	Discharge in 2004	Target in 2009
Chiba Prefecture	t day^{-1}	51	46	42	36	45	43	36	33	3.4	3.0	2.7	2.3
Tokyo Metropolitan	t day^{-1}	73	70	61	58	101	100	78	76	7.7	7.2	6.0	5.8
Kanagawa Prefecture	t day^{-1}	30	28	27	26	42	41	33	31	3.5	3.2	2.5	2.0
Saitama Prefecture	t day^{-1}	93	84	81	73	66	65	61	59	6.5	5.8	4.1	3.8
Total of 4 local gov.	t day^{-1}	247	228	211	193	254	249	208	199	21.1	19.2	15.3	13.9
Reduction in 1999–2004	t day^{-1}		19				5				1.9		
Reduction in 1999–2004	t year^{-1}		6900				1800				690		
Reduction in 2004–2009	t day^{-1}				18				9				1.4
Reduction in 2004–2009	t year^{-1}				6600				3300				510

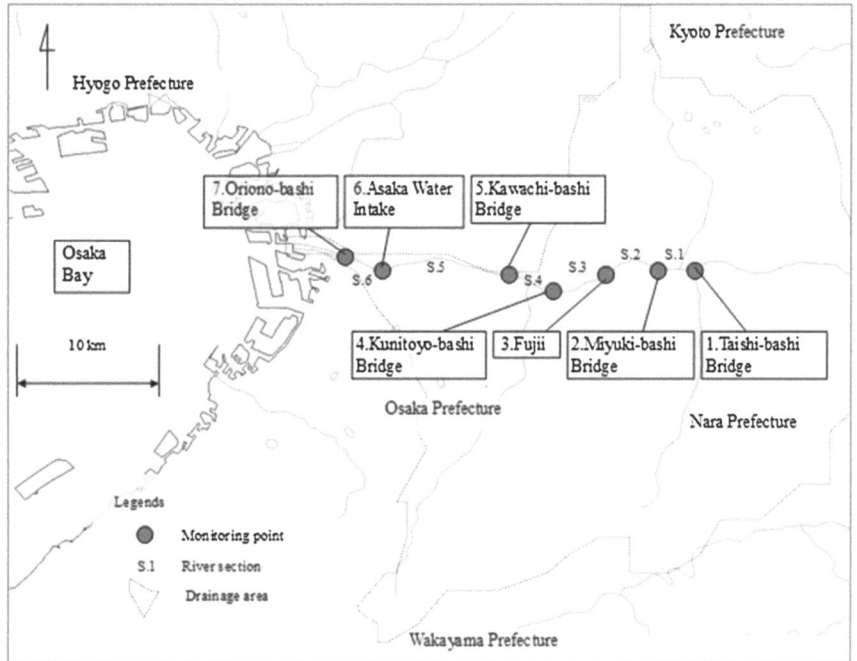

Fig. 2.6 Yamato-gawa River Basin, Japan (Tsuzuki et al. 2010a). (Copyright permission has been obtained from Elsevier)

It has been also observed that the relationship between pollutant discharge and water quality has not in accordance with a linear relationship in a short-term rainfall events in a mountainous catchment in Kyoto-Fu, Japan (Fujii et al. 2006), and in the Richmond River, Australia (McKee et al. 2000). The short-term relationship between pollutant discharge and water quality which is not in accordance with a linear relationship is called hysteresis.

2.4   Natural Purification of BOD in the River Sections

Natural purification effects in ambient water including the rivers are generally well known. For example, a classical relationship between BOD and DO has been proposed (Eqs. 2.6 and 2.7) (Streeter and Phelps 1925; Toda 2001; Tsuzuki et al. 2010a).

$$\text{BOD} : u\frac{dB}{dx} = -k_bB - k_pB + L_B \tag{2.6}$$

$$\text{DO} : u\frac{dD}{dx} = -k_bB + k_r(D^* - D) - L_D \tag{2.7}$$

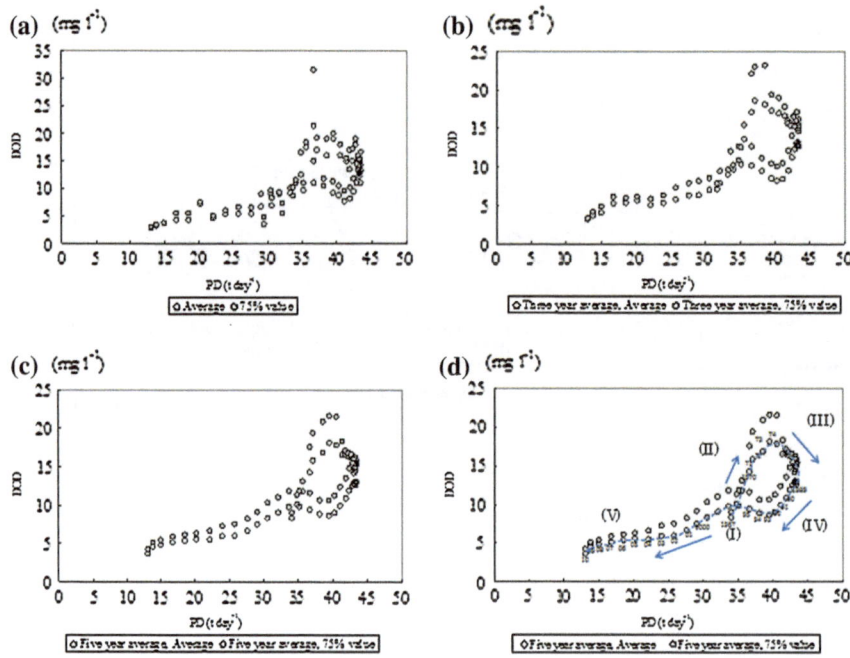

Fig. 2.7 Chronological relationship between BOD discharge in the Yamato-gawa River Basin and BOD quality in the river, **a** average and 75 percentile, **b** 3 year average of average and 75 percentile, **c** 5 year average of average and 75 percentile, and **d** 5 year average of average and 75 percentile with explanation of Phases (Tsuzuki et al. 2013). (Copyright permission has been obtained from Springer)

where B is BOD concentration (g m^{-3}); D is DO concentration (g m^{-3}); x is distance (m); u is advection velocity (m s^{-1}); k_b is biological oxygen consumption rate (min^{-1}); k_p is BOD removal rate with physical and chemical reaction (s^{-1}); L_B is BOD loading (g m^{-3} s^{-1}); k_r is re-aeration coefficient (s^{-1}); D^* is saturation oxygen concentration (g m^{-3}); and L_D is oxygen consumption rate with other reaction than biological (g m^{-3} s^{-1}).

The Streeter-Phelps equation has been applied to BOD and DO concentrations in the six sections of the Yamato-gawa River, and the values of k_b, biological oxygen consumption rate, and k_p, BOD removal rate with physical and chemical reaction have been investigated to evaluate the magnitude of natural purification effects of these river sections (Tsuzuki et al. 2010a). The magnitude of natural purification effects has been in accordance with the existing research (Kusuda 1986; Shimomura et al. 2008). The effects the soft measures in households have been quantified based on the results (Fig. 2.8). When all the households in the river basin are assumed to introduce and continue the soft measures, BOD concentration at the monitoring point nearest to the river mouth has been estimated to improve 25 %, from 4.1 to 3.1 mg-BOD l^{-1}.

Fig. 2.8 BOD and DO estimation results of the Streeter-Phelps one dimensional water quality model with and without soft measures in households. (Modified from Tsuzuki et al. 2010a). (Copyright permission has been obtained from Elsevier)

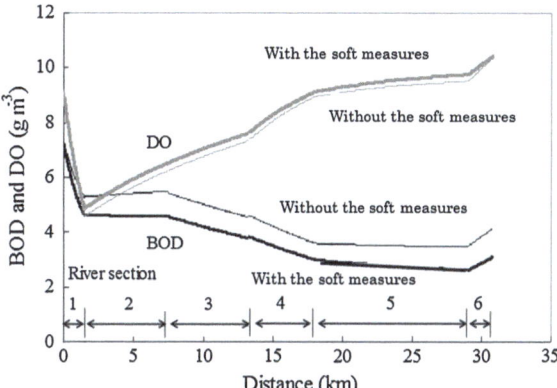

Fig. 2.9 Pollutant runoff yield from corresponding sub-river basin, $Rd_{(n+1)(n+1)}$, and that from upper monitoring point, $Rm_{n(n+1)}$ (Tsuzuki and Yoneda 2011). (Copyright permission has been obtained from Elsevier)

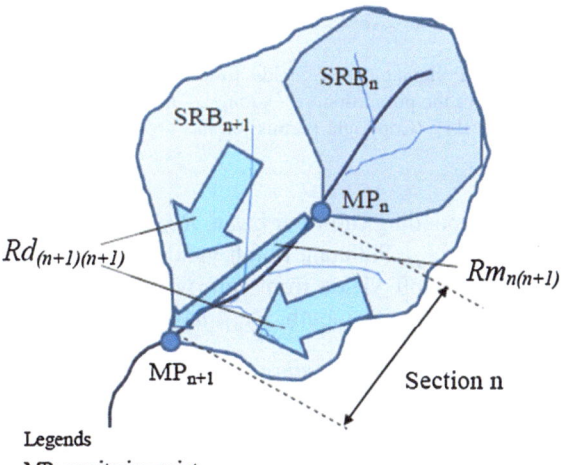

Legends
MP: monitoring point
$Rd_{(n+1)(n+1)}$: runoff yield from the corresponding sub-river basin
$Rm_{n(n+1)}$: runoff yield from the upper monitoring point
SRB: sub-river basin

2.5 Pollutant Runoff Yields in Sub-Catchments

Some parts of the pollutants discharged to ambient water are purified and cleaned in the ecosystems, e.g. being uptaken biologically by organisms including fish, plankton, invertebrates and microorganisms, and physically settling to the bottom of the rivers, lakes and coastal areas, or attaching to particles. Pollutant runoff yield is a coefficient to indicate the ratio of pollutant load in the river and pollutant discharged to the catchment or river basin (Fig. 2.9) (Tsuzuki and Yoneda 2011). Pollutant runoff yield at monitoring point, MP_{n+1}, has been attributed to that from

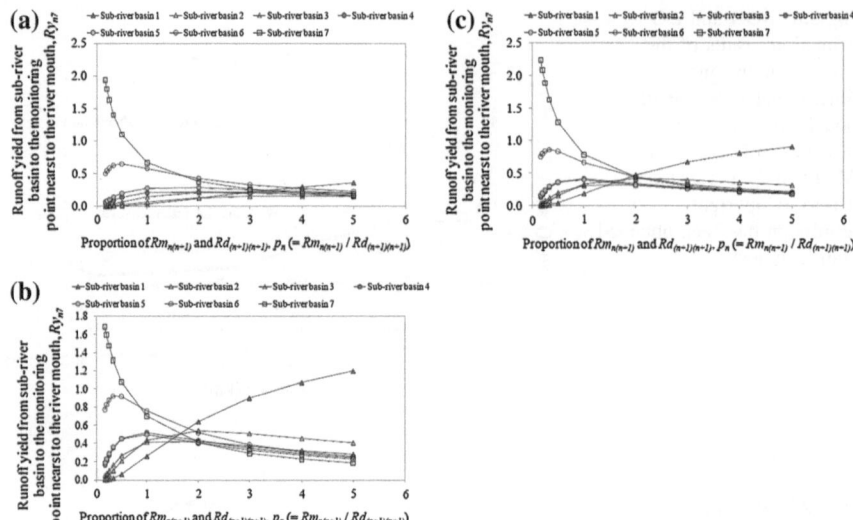

Fig. 2.10 Pollutant runoff yields from sub-river basins n to monitoring point 7, Ry_{n7}, as functions of the proportion, p_n ($= Rm_{n(n+1)} / Rd_{(n+1)(n+1)}$) for **a** BOD **b** TN and **c** TP (Tsuzuki and Yoneda 2011). (Copyright permission has been obtained from Elsevier)

the corresponding sub-river basin, $Rd_{(n+1)(n+1)}$, and that from the upper monitoring point, $Rm_{n(n+1)}$. Pollutant runoff yield from MP_n to MP_{n+1} is described as Eq. 2.8. Pollutant runoff yields from sub-river basins (from 1 to 7) to monitoring point nearest to the river mouth, Ry_{n7}, have been calculated with Eq. 2.9.

$$Rm_{n(n+1)} = \frac{(POLloading)_{MP(n+1)} - (POLdisch)_{SRB(n+1)} \times Rd_{(n+1)(n+1)}}{(POLloading)_{MP(n)}} \quad (2.8)$$

where $Rm_{n(n+1)}$ is runoff yield from monitoring point MPn to MPn+1 (%) (POL-loading)$MP_{(n+1)}$ is pollutant loading at monitoring point MP_{n+1} (kg day^{-1}) (POLdisch)$SRB_{(n+1)}$ is pollutant discharge in sub-river basin MP_{n+1} (kg day^{-1}); $Rd_{(n+1)(n+1)}$ is pollutant runoff yield from sub-river basin n+1 to MPn+1 (−).

$$Ry_{n7} = Rd_m \times Rm_{n(n+1)} \times \cdots \times Rm_{n7} \ (1 \leq n \leq 6)$$
$$= Rd_{77} \quad\quad\quad\quad\quad (n = 7) \quad\quad (2.9)$$

where Ry_{n7} is pollutant runoff yield from sub-river basin n to monitoring point, MP_7 (%).

Pollutant runoff yields from sub-river basins to the monitoring point 7, Ry_{n7}, are expressed as functions of the proportion, p_n (Fig. 2.10). P_n is the ratio of $Rm_{n(n+1)}$ and $Rd_{(n+1)(n+1)}$ (Eq. 2.10). The results of BOD show Ry_{n7} increases with pn increase when pn is small in sub-river basins 1–6. When p_n is larger than the corresponding value to the each curve peak of Ry_{n7}, Ry_{n7} decreases with p_n increase. The corresponding values of Ry_{n7} peaks for BOD are 4.0 for sub-river basin 2, 3.0 for sub-river basin 3, 2.0 for sub-river basins 4 and 5, and 0.5 for sub-river basin 6 (Fig. 2.10a). Rr_{17} always increases when p_n is from 0.17 to 5.0. On the contrary, for sub-river basin 7, pollutant runoff yield from sub-river basin 7, Ry_{77} (= Rd_{77}), decreases with p_n increase. Tendencies of Ry_{77} are the same for TN and TP besides the corresponding values for Ry_{n7} curves are different by pollutant (Fig. 2.10b and c).

$$P_n = \frac{Rm_{n(n+1)}}{Rd_{(n+1)(n+1)}} \tag{2.10}$$

References

Fujii S, Shivakoti BR, Shichi K, Songprasert P, Ihara H, Moriya M, Kitpati S, Tanaka S (2006) Analysis of parameter variations in L-Q equations for river runoff processes from the viewpoint of spatial and temporal conditions. Water Sci Technol 53(10):141–152

Kusuda T (1986) Purification mechanisms in the river, Japan Society of Civil Engineers Sanitary Engineering Committee Report of Attached Microorganisms, 95–115 (in Japanese). *cited in* Kunimatsu, T. and Muraoka, K. (1989) *Model analysis of river pollution*, Gihodo, Tokyo, Japan, 266p. (in Japanese)

McKee L, Eyre B, Hossain S (2000) Intra- and inter annual export of nitrogen and phosphorus in the subtropical Richmond River catchment. Hydrolo Process 14:1787–1809

Shimomura T, Ii H, Taniguchi M (2008) Effect of the river purification facilities in the Yamato River. In: Proceedings of the annual meeting of environment and hydrology group of hydrology committee, Japan Society of Civil Engineers. (in Japanese)

Streeter HW, Phelps EB (1925) Studies of the pollution and natural purification of the Ohio River, Part III, Factors concerned in the phenomena of oxidation and reaeration, Public Health Bulletin No. 146. U.S. Public Health Service, Washington

Toda Y, (2001) One dimensional analysis of BOD and DO in the river. In: Hydraulics formulae: hydraulics worked examples with CD-ROM, Japan Society of Civil Engineers Hydraulics Committee. (in Japanese)

Tsuzuki Y (2004) Land based water pollutant loads of Tokyo Bay and nutrients removal by fishery, environment system. In: Proceedings of 32nd annual meeting of environmental systems research. pp 413–418. (in Japanese with English abstract)

Tsuzuki Y (2005) Environmental accounting housekeeping (EAH) books of domestic wastewater. Water Supply Wastewater Treat (Yousui to Haisui) 47(7):539–545 (2005). (in Japanese)

Tsuzuki Y (2006) An index directly indicates land-based pollutant load contributions of domestic wastewater to the water pollution and its application. Sci Total Environ 370(2–3):425–440

Tsuzuki Y (2013) Explanation of 47-Year BOD Alternation in a Japanese River Basin by BOD Generation and Discharge, Water, Air, & Soil Pollution, 224 (5), (Article 1517)

Tsuzuki Y, Koottatep T, Wattanachira S, Sarathai Y, Wongburana C (2009) On-site treatment systems in the wastewater treatment plants (WWTPs) service areas in Thailand: Scenario based pollutant loads estimation. J Glob Environ Eng Jpn Soc Civil Eng 14:57–65

Tsuzuki Y, Fujii M, Mochihara Y, Matsuda Y, Yoneda M (2010a) Natural purification effects in the river in consideration with domestic wastewater pollutant discharge reduction effects. J Environ Sci 22(6):892–897

Tsuzuki Y, Koottatep T, Jiawkok S, Saenpeng S (2010b) Municipal wastewater characteristics in Thailand and effects of "soft intervention" measures in households on pollutant discharge reduction. Water Sci Technol 62(2):231–244

Tsuzuki Y, Yoneda M (2011) Pollutant runoff yields in the Yamato-gawa River, Japan, to be applied for EAH books of municipal wastewater intending pollutant discharge reduction. J Hydrol 400(3–4):465–476

Tsuzuki Y, Koottatep T, Sinsupan T, Jiawkok S, Wongburana C, Wattanachira S, Sarathai Y (2013a) A concept in planning and management of on-site and centralised municipal wastewater treatment systems, a case study in Bangkok, Thailand I: pollutant discharge indicators and pollutant removal efficiency functions.Water Sci Technol 67(9):1922–1933

Chapter 3
Soft Measures in Households

Soft measures in households[1] are measures which can be conducted in households to decrease wastewater pollutant discharge amounts. A pollutant discharge calculator is of municipal wastewater[2] is a tool to estimate pollutant discharge reduction effects by soft measures. This is an effective tool for citizens to estimate municipal wastewater pollutant discharge and pollutant load reductions by soft interventions in households and also for scientific purposes. The pollutant discharge calculator of municipal wastewater has been derived from the pollutant discharge calculator of carbon dioxide (CO_2) or CO_2 footprint calculator. The CO_2 or carbon footprint calculators are more popular tools in the world, which have been prepared by national and local governments, utilities, environment NGOs and universities as countermeasures to the global warming in order to estimate and to reduce CO_2 emissions from municipal lives.

Dissemination of water environment information to ordinary people is necessary to improve community participation in ambient water quality improvement or to increase proliferation rates of soft measures in households. Stakeholders of water environment including researchers, administrations, professionals, industries, environmental NGOs, fishermen and ordinary people have their roles. The pollutant discharge calculator of municipal wastewater will be one of the tools for enhancing community participation. Estimation results of pollutant removals by fishery in Tokyo Bay are also described to consider pollutants in the enclosed coastal sea.

---

[1] "Soft measures" are measures which are conducted in daily life of ordinary people especially in households to reduce municipal wastewater pollutant discharge. On the contrary, hard measures are facilities or systems to reduce pollutant load in ambient water including on-site and centralised wastewater treatment systems and river water purification facilities. Soft measures are formerly named as soft interventions.

[2] An environmental accounting housekeeping (EAH) book of municipal wastewater is a former name of this tool (Tsuzuki 2006).

Y. Tsuzuki, *Pollutant Discharge and Water Quality in Urbanisation*,
SpringerBriefs in Water Science and Technology,
DOI: 10.1007/978-3-319-04756-0_3, © The Author(s) 2014

3.1 Soft Measures in Households and Their Dissemination: A Social Experiment Programme in the Yamato-gawa River Basin[3]

Municipal wastewater pollutant discharge reduction and river water quality improvement have been pursued in the Yamato-gawa River basin, Japan, since 2005 with a community participation program called the Yamato-gawa River Social Experiment Program for reducing domestic wastewater pollutant discharge (Tsuzuki 2010b; Tsuzuki et al. 2009b, 2012). The Social Experiment Program has been conducted in the framework of the C-Project, which aims to "change" the river image of deteriorated water quality, "collaborate" in all the river basin, and to "concentrate" into the project (Yamato-gawa River Office, 2006–2010). The Social Experiment Program is the first community participation program aiming to reduce municipal wastewater pollutant discharge in the river basin scale of about 2 million population, and maybe in the world. The Yamato-gawa River and its branch has been nominated as one of the worst water quality rivers in Japan in these decades judging from larger average BOD concentration at the monitoring points.

In the Social Experiment Program, such kinds of information as soft inter-ventions in households to reduce municipal wastewater pollutant discharge have been disseminated to ordinary citizens in the river basin. The three major soft interventions in the Programme are (1) to prepare foods and drinks only the amount which you eat and drink (2) to stop draining residual liquids and foods and put them into solid wastes, and (3) to wipe dish and cooking implements or cookware up with paper or rag before washing them in kitchen (Table 3.1).

Dissemination of the soft interventions in households raises people's awareness about the relationship between pollutant discharges from their daily lives and river water environment. We hope our experiences in the Yamato-gawa River basin will also contribute to water quality improvement in other water bodies in the world (Tsuzuki et al. 2010b, 2012).

Table 3.1 shows soft measures or soft interventions in households which have been disseminated to ordinary people in the Social Experiment Program (SEP) in the Yamato-gawa River Basin since 2005 (Tsuzuki et al. 2012). Soft measures A1 and A2 relate to attitudes on foods and drinks. Behaviours in the kitchen are B1–B6. C1 is on the dining. D1–D3 also relate to the kitchen and dining and focusing treatment after eating. E1–E3 relate to behaviours in washing clothes. F1 relate to on-site wastewater treatment facilities. Effects of soft and hard interventions on BOD discharge/load reduction in the main channel of the Yamato-gawa River are estimated based on BOD discharge/load in 2005 (Fig. 3.1a). The effects of soft interventions on pollutant discharge/load reduction are found to be comparable to

---

[3] This section is modified from International Water Association (IWA) Specialist Group on Diffuse Pollution Newsletter, No. 31, September, 2010 (Tsuzuki 2010b).

Table 3.1 Soft measures (soft interventions) in households to reduce pollutant discharge (modified from Tsuzuki et al. 2012) (Copyright permission has been obtained from Elsevier)

Soft intervention	ΔPDC[a] (%)	Proliferation rate[b]
(A1) To prepare foods only the amount to eat and not to waste residue	5	(I)
(A2) Not to drain liquid residues including *miso* soup, noodle soup and beer	5	(I)
(B1) To throw residual foods and liquids as solid wastes rather than drain as wastewater	5	(I)
(B2) To absorb residual liquid with used paper or rug and to waste them away as solid wastes	5	(III)
(B3) To install a plastic triangle corner and a basket in the kitchen	3	(I)
(B4) To attache plastic bags/nets or paper filter to a plastic triangle corner and a basket	3	(I)
(B5) To pour rice washing water to vegetables or plants	6	(III)
(B6) To make residue oil solid wastes or to be collected by local governments and so on	5	(II)
(C1) To wipe dish and cooking apparatus up before washing	10	(II)
(D1) To decrease detergent usage amount	3	(I)
(D2) To use acrylic sponge in the kitchen	2	(II)
(D3) To use detergent with less environmental burden including sodium bicarbonate	3	(II)
(E1) To decrease detergent for washing clothes	4	(II)
(E2) To recycle bath tab water for washing clothes	4	(II)
(E3) To take measure of the amount of detergent used for washing clothes	1	(I)
(F1) To properly understand and make use of *johkasou*, on-site wastewater treatment system	NA[c]	(III)

[a] Pollutant discharge reduction effect of soft measures
[b] Proliferation rate at seventh SEP in Feb 2010 except for (E3) at fifth SEP in Feb 2008 because of data availability (I): >80 % (II): 60–80% (III): <60 %
[c] not available

those of hard intervention in 2007–2009. Hard interventions mean development of municipal wastewater treatment systems and river water purification facilities. The soft interventions are found to be significantly cost effective in reducing pollutant discharge/load in the river basin (Fig. 3.1b). Effects of soft and hard interventions on pollutant discharge/load are considered to be benefits of these measures in the figure, and benefit-cost ratio is compared.

3.2 Stakeholders on Water Environment

Centralized wastewater treatment systems are usually developed and managed by the administrations, whereas on-site wastewater treatment systems are usually installed and periodically maintained by owners of houses or buildings using the

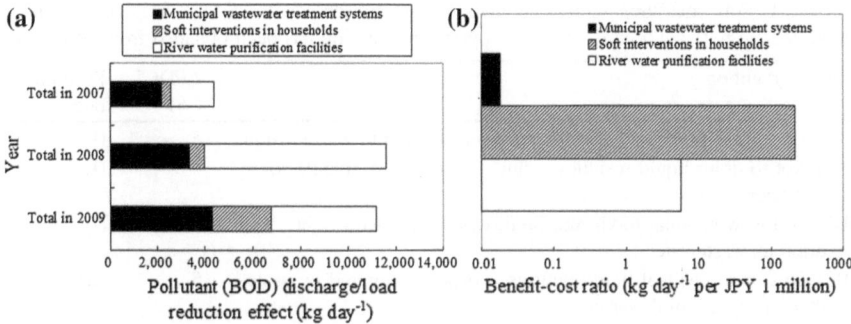

Fig. 3.1 Comparison of soft interventions and hard interventions (development of municipal wastewater treatment systems and river water purification facilities), **a** pollutant discharge/load reduction effects, and **b** benefit-cost ratio (Tsuzuki et al. 2012) (Copyright permission has been obtained from Elsevier)

on-site system companies. There are also subsidies or financial assistances for the on-site wastewater treatment systems, CJ, to enhance development of on-site treatment systems and improvement of on-site treatment systems from the old-type treatment systems (SJ and NST) to new-type treatment systems (CJ). Dissemination of municipal wastewater treatment systems with their performances, i.e. pollutant discharge removal efficiencies and PDCs may enhance the understanding of these on-site wastewater treatment systems and the relationship between pollutant discharge and ambient water quality by the ordinary people (Fig. 3.2). Stakeholders of marine water pollution problem in coastal areas are summarized as Table 3.2 (Tsuzuki 2004). Other stakeholders in Fig. 3.2 include industries and agriculture.

These kinds of participatory approaches have been already conducted in some projects and programs in developing countries (Fig. 3.3). PROSANEAR (National sewerage strategy project) is an experience of Brazil. More than 100 communities and 17 cities participated in the PROSANEAR project or the Low Income Sanitation Technical Assistance Project—PROSANEAR—TAL (World Bank 2008). Total population in these communities and cities has been ca. 900,000. Condominal sewerage has been adopted in 9 cities. Stabilization ponds, septic tanks, and upflow anaerobic sludge blanket (UASB) processes have been adopted for sewerage treatment. Total investment amount has been US$91.5 million. Per capita cost for sewerage development has been US$104 person^{-1} (Katakura and Bakalian 1998).[4] A community selected one option from a range of options for connection to a public sewerage system, which have been provided by the implementing agency, through a process of discussion among the community residents. The project has been approved in 2000 and completed in 2007.

---

[4] The investment amount in the literature is US$91.5 billion, however, it may be US$91.5 million based on the estimation from the other World Bank report (World Bank 2008).

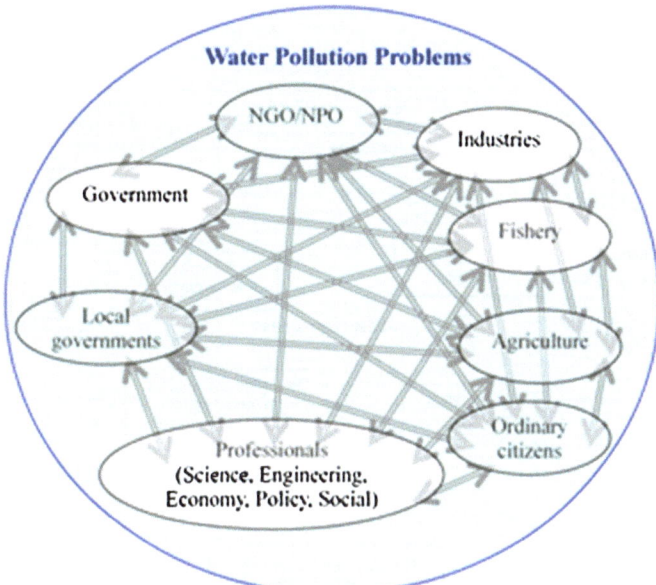

Fig. 3.2 Stakeholders of water pollution problems. Dissemination and environmental education for ordinary citizens and dialogues between stakeholders including fishery and agriculture are desirable. (Modified from Tsuzuki 2006) (Copyright permissions have been obtained from Elsevier and Nova Science Publishers)

Table 3.2 Stakeholders of water pollution problems in Tokyo Bay

Stakeholder	Role
Researcher	Monitoring and dissemination of environment data
	Theoretical research
Research institute	Monitoring and dissemination of environment data (Theoretical research)
Administration	Monitoring and dissemination of environment data
	Planning and enforcement of policy and interventions
	Adjustment among multiple administrations
Fisherman	Fishery (Information on fish, sea shell and sea weed)
Environment	Environmental related activities
NGO/NPO	Connection between administration, researchers and ordinary people
Ordinary people	Eating fishery products (with judgment of toxic)

(Modified from Tsuzuki 2004)

After the PROSANEAR project in Brazil, the World Bank conducted similar wastewater treatment projects in the Latin America and Caribbean regions.

Disseminations on water quality and pollutant discharges have been conducted by the national and local governments and environmental NGOs in Japan since the 1980s. For examples, some most prevailing dissemination contents have been desirable usage methods of kitchen, decreasing usage amount of detergents in the

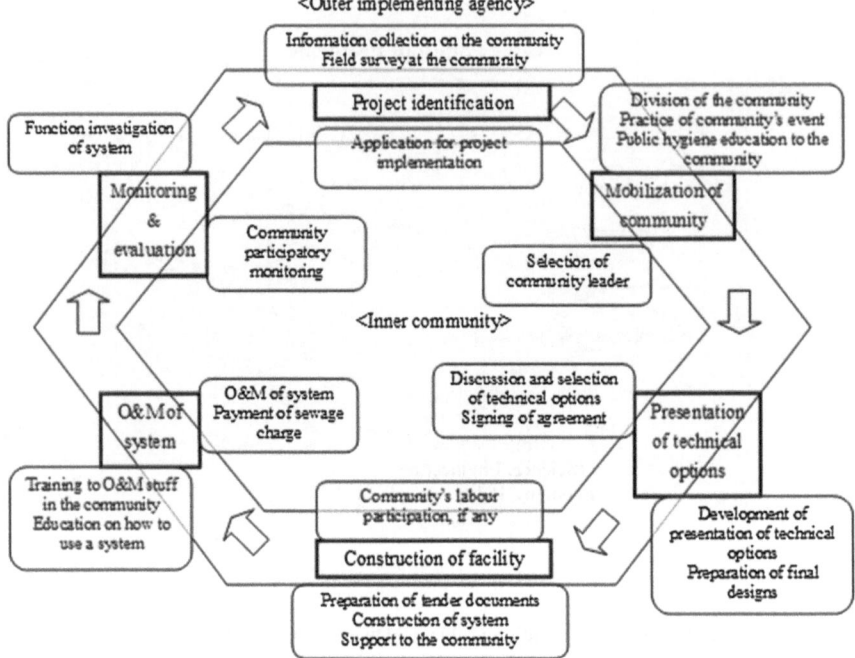

Source: Modified from Infrastructure Development Institute, Japan (2004)

Fig. 3.3 Responsibility sharing between the implementing agency and the community. An example of PROSANEAR in Brazil. (Modified from IDIJ 2004) (Copyright permissions have been obtained from IDIJ and Nova Science Publishers)

kitchen and bath, and how many corresponding bath tab water or 300 L of water amounts are required to dilute the wastewater discharges, e.g. residue of noodle soup of 100 ml, to the water quality in which fish can survive (Table 3.3). Table 3.3 is prepared using Tokyo Metropolitan (2008), which is based on Tokyo Metropolitan (1986). These dissemination contents have been summarised into and improved to the soft interventions or soft measures in the Social Experiment Program in the Yamato-gawa River Basin.

3.3 Dissemination to Ordinary Citizens: Water Pollutant Discharge and CO_2 Emission

Contributions of municipal wastewater discharges to ambient water quality can be estimated by pollutant discharge calculators of municipal wastewater (see Chap. 2; Table 2.1) and PLC_{wb} by river sub-basin and by wastewater treatment method (Fig. 2.2). Pollutant discharge calculator of municipal wastewater (Table 2.1) can

Table 3.3 How many bath tab water is needed to dilute up to water quality which is suitable for fish to live in (BOD is 5 mg l^{-1})? (Prepared based on Tokyo Metropolitan 2008)

Pollutant	Volume (ml)	BOD (ms l^{-1})	How many bath tab water is needed to dilute to 5 mg l^{-1}?[a]
Sov source	15	150,000	1.5
Miso soup	200	35,000	4.7
Oden Soup	500	74,000	25
Used tenpura oil	500	1,000,000	330
Rice washing water	2,000	3,000	4
Milk	200	78,000	10
Chinese noodle soup	200	25,000	3.3
Japanese wine	20	200,000	2.7
Beer	180	81,000	17

[a] Bath tab volume is supposed to be 300 L

be utilized for dissemination and estimation of pollutant discharges and their reduction amounts by conducting soft interventions in households.

Pollutant discharge calculator of carbon dioxide (CO_2) emission reduction or carbon footprint calculators as countermeasures against climate change are widely developed by the national and local governments, utilities, environmental NGOs and universities in Japan (Table 3.4).

The concept of EAH book or '*Kankyo Kakeibo (in Japanese)*' has been firstly proposed by Morioka (1986) (Takeda 2005; Tsuzuki and Yoneda 2011; Tsuzuki 2014). For example, in the backside of monthly bill from Tokyo Electric Power Co. Ltd. (TEPCO), a table to estimate CO_2 emissions in households has been provided by May 2011 (see Tsuzuki 2010a), which has been stopped because of the company conditions changed after the Earthquake disaster in March 2011. Basic structure of the EAH book of CO_2 emission is shown in Table 3.5. People can calculate CO_2 emission and consumption costs of several daily activities using coefficients. Coefficients are determined by area and by time (year 2010, 2011, month etc.).

Recently, the number of websites on CO_2 emission reduction or carbon footprint calculaters has also increased in other countries than Japan, including the UK, USA and Australia (Table 3.4). Padgett et al. (2008) compared several websites on carbon emission calculation.

3.4 Pollutant Discharge Calculators of Municipal Wastewater

Pollutant discharge reduction has been estimated to be 38–53 % for BOD, TN and TP, which has been estimated using the pollutant discharge calculator of municipal wastewater, a spreadsheet type pollutant discharge reduction estimation tool

Table 3.4 Websites of the pollutant discharge calculators of CO_2 emission in Japan and abroad

Type of organisation	Organisation	URL	Note
National government	Ministry of the environment	http://www.eco-family.go.jp/	Eko Cho (Ecology Notebook)' organised from 2004 to 2011 in the programme of 'Environmental Minister in my family'
		http://www.env.go.jp/policy/wagaya/ http://www.eeel.go.jp/kakeibo/	
Third sector	Environmental Information and Communication Network	http://www.eic.or.jp/ecoterm/?act= view&serial=482	Environmental Household Bookkeeping System/Household Eco-Account Book
Local government	Chiba Prefecture (1)	http://www.wit.pref.chiba.lg.jp/slim/ slim2.htm	Web-based EAH books of CO_2. Organised by Environment Research Centre
	Chiba Prefecture (2)	http://www.pref.chiba.lg.jp/kansei/ kankyougakushuu/fukudokuhon/ gakushuu-36.html	Checklist of daily lives. downloadable PDF file. Its concept is similar to the original EAH book by Morioka (1986)
	Kagawa Prefecture	http://www.pref.kagawa.jp/kankyo/ chikyu/kakeibo/index.htm	Downloadable MS-Excel file. support was completed in 2004
	Yamanashi Prefecture	http://www.pref.yamanashi.jp/ kankyo-sozo/kakeibo.html	Downloadable PDF file
	Kyoto City	http://www.doyoukyoto.com/	Web-based EAH books of CO_2. Need registration
	Fukuoka City	http://www.ecofukuoka.jp/	Web-based EAH books of CO_2. Need registration
	Sakura City	http://www.city.sakura.lg.jp/ 012543000_kankyohozon/osirase/ setsuden.html	Downloadable PDF file

(continued)

Table 3.4 (continued)

Type of organisation	Organisation	URL	Note
	Yawatahama City	http://www.city.yawatahama.ehime.jp/03jyouhou/kankyo/kakeibo/kankyoukakeibo.htm	Downloadable MS-Excel file and PDF file
	Yokohama City	http://www.city.yokohama.lg.jp/kankyo/ondan/ecohama/	Web-based EAH books of CO_2. Need registration
Environmental NGO	CASA	http://www.shiftra.jp/casa/system/indexphp?&ac=what	Web-based EAH books of CO_2. Need registration
	Network Earth Village ('Chikyu-mura')	https://www.stop-ondanka.com/	Web-based EAH books of CO_2. Need registration
	Environment Network in Yamagata	http://eny.jp/eco_account_book/about_challenger.html	Web-based EAH books of CO_2. Need registration
	Environment LOHAS	http://www.lohasclub.org/carbonfree/200.html	Coefficients of electricity, gas and water are summarised at prefecture-based
University	Miyagi Education University	http://eml.edb.miyakyo-u.ac.jp/JOHO/kakeibo/kids/index.htm	Web-based EAH books of CO_2
	Kyoto Gakuen University	http://biseibutsu.exblog.jp/5033301/	List of EAH books of CO2 in 2007. Prepared independently by yshinoda
Private company	Panasonic	http://panasonic.co.jp/phc/eco/cm/03.html	Promoting EAH book of CO_2 to employees and their family
	Tokyo Gas, Saitama branch	http://www.tokyo-gas.co.jp/area/saitama/environment/ecolifeday.html	Social programme to reduce CO_2 emission by use of the Ecology life checksheet
	Aeon Integrated Business Service	https://www.ecohana.jp/ecosaas/ecohana.html	Links shopping to CO2 emissions. This service has been conducted as a project of the Ministry of Internal Affairs and Communications

(continued)

Table 3.4 (continued)

Type of organisation	Organisation	URL	Note
Others	Naver matome	http://matome.naver.jp/odai/2128832004346101	Links to the websites where you can find information on 'Kankyo Kakeibo' or download its files
Other countries than Japan			
Australia	Michael Bloch and Carbonify.com	http://www.carbonify.com/carbon-calculator.htm	Carbon dioxide emission footprint calculator and offset estimator
UK	Carbon Footprint Ltd	http://www.carbonfootprint.com/termsandconditions.html	Calculator of carbon footprint and carbon offset
UK	Directgov, the UK government's digital service for people in England and Wales	http://www.direct.gov.uk/en/Dio11/DoItOnline/DG_10015994	Do it online: Calculate your car's tax and CO_2 emissions
USA	The Nature Conservancy	http://www.nature.org/greenliving/carboncalculator/index.htm	Carbon footprint calculator
USA	USEPA	http://www.epa.gov/climatechange/emissions/ind_calculator.html	Household emissions calculator
USA	USEPA	http://www.epa.gov/cleanenergy/energy-resources/calculator.html	Greenhouse gas equivalencies calculator

There are also other names including EAH books and Household Eco-Account Books

Table 3.5 Basic structure of the EAH book of CO_2 or carbon footprint calculators

	Consumption amount	Unit	Coefficient	CO_2 emission	Cost
Electricity		kWh	0.43		
Gas (piped)		m^3	2.28		
LP gas		m^3	6.22		
Petroleum		L	2.49		
Gasoline (for car)		L	2.32		

(Table 2.1) (Tsuzuki 2006; Tsuzuki et al. 2009a, 2010b). These soft interventions have been mixed with hard interventions, i.e. development of wastewater treatment facilities and river water purification facilities. The latter is installed in the river bed to improve river water quality. Large natural purification effects in the river section with the river water purification facilities have been evaluated by use of biological oxygen consumption rate estimation (Tsuzuki et al. 2010a).

References

Katakura Y, Bakalian A (1998) PROSANEAR: people, poverty and pipes, UNDP-World Bank, Water and Sanitation Program. (World Bank website)

Morioka T (1986) Building up close-at-home environment: EAH book of CO_2 and environment chart (Mijikana Kankyou Dukuri: Kankyou Kakeibo to Kankyou Karute). Nihon-Hyoron-Sha Publishing, p 255 (in Japanese)

Padgett JP, Steinemann AC, Clarke JH, Vandenbergh MP (2008) A comparison of carbon calculators. Environ Impact Assess Rev 28:106–115

Takeda H (2005): Governance through the family: the political function of the domestic in Japan. In: Hook GD (ed) Contested governance in Japan (Chap. 12): sites and issues, vol 265. Routledge Curzon, Oxon and New York, p 233–245, ISBN: 0-415-36498-1

Tokyo Metropolitan (2008) Improvement of water environment in the rivers and coastal areas (in Japanese). http://www.kankyo.metro.tokyo.jp/water/attachement/H19-panf.pdf

Tokyo Metropolitan (1986) Guidelines of municipal wastewater treatment. (in Japanese)

Tsuzuki Y (2004) Land based water pollutant loads of Tokyo Bay and nutrients removal by fishery, Environment System. In: Proceedings of 32nd annual meeting of environmental systems research, pp 413–418. (in Japanese with English abstract)

Tsuzuki Y (2006) An index directly indicates land-based pollutant load contributions of domestic wastewater to the water pollution and its application. Sci Total Environ 370(2–3):425–440

Tsuzuki Y (2010a): Domestic wastewater pollutant discharge and pollutant load water quality in the ambient water in developed and developing countries (Chap. 5). In: Kudret Ertuð and Ilker Mirza (eds) Water quality: physical, chemical and biological characteristics, vol 277. Nova Science Publishers Inc., New York, pp 125–164, ISBN: 978-1-60741-633-3

Tsuzuki Y (2010b). Municipal wastewater pollutant discharge reduction with community participation in the Yamato-gawa River Basin, Japan, International Water Association (IWA) Specialist Group on Diffuse Pollution Newsletter No. 31, September 2010. http://www.iwahq. org/Home/Networks/Specialist_groups/List_of_groups/Diffuse_Pollution

Tsuzuki Y (2014) Evaluation of the soft measures' effects on ambient water quality improvement and household and industry economies, J Clean Prod 66:577–587

Tsuzuki Y, Yoneda M (2011) Pollutant runoff yields in the Yamato-gawa River Japan, to be applied for EAH books of municipal wastewater intending pollutant discharge reduction. J Hydrol 400(3–4):465–476

Tsuzuki Y, Koottatep T, Rahman MM (2009a) Water quality profiles of the tidal rivers and canal in per-urban of Bangkok Thailand, and Dhaka, Bangladesh, focusing on the water quality transition in coastal areas. J Japan Society Water Environ 32(1):47–52. (in Japanese with English abstract)

Tsuzuki Y, Fujii M, Mochihara Y, Matsuda Y, Yoneda M (2009b) Domestic wastewater pollutant discharge reduction with the community participation program in the Yamato-gawa River drainage area, Japan, 12th International Riversymposium, Australia, 21–24 Sep 2009

Tsuzuki Y, Fujii M, Mochihara Y, Matsuda Y, Yoneda M (2010a) Natural purification effects in the river in consideration with domestic wastewater pollutant discharge reduction effects. J Environ Sci 22(6):892–897

Tsuzuki Y, Koottatep T, Jiawkok S, Saenpeng S (2010b) Municipal wastewater characteristics in Thailand and effects of "soft intervention" measures in households on pollutant discharge reduction. Water Sci Technol 62(2):231–244

Tsuzuki Y, Yoneda M, Tokunaga R, Morisawa S (2012) Quantitative evaluation of effects of the soft interventions or cleaner production in households and the hard interventions: A Social Experiment Programme in a large river basin in Japan. Ecol Ind 20:282–294

World Bank (2008) Low Income Sanitation Technical Assistance Project—Prosanear—TAL. (World Bank website)

Chapter 4
Relationship Between Economic Development and Pollutant Discharge per Capita (PDC)

Relationship between economic development and pollutant discharge per capita (PDC) has been inverted U-shaped curve, namely environmental Kuznets curve (EKC), in Japan and developing countries. PDC-COD_{Mn}, PDC-TN and PDC-TP increased after 1940s in the economic development after the World War II, and PDC decreased after 1970 when a series of environmental regulations have been established in the drainage area of Lakes Shinji and Nakaumi, Japan (Tsuzuki 2007, 2009a). The relationship between water quality and economic indicators is often regarded as inverted U-shaped curve or EKC (Grossman and Krueger 1995). The relationship between economic development and PDC of all the developing countries has not been similar to the EKC, however, the relationships of only the countries in Asia, Pacific and Africa regions have been found to be similar to the original Kuznets relationship of income and inequity, i.e. income inequity is large when income is small, and income inequity is small when income increases to some extent (Tsuzuki 2009c). PDC has been in a wide range in the countries with smaller income, and PDC has been in a definite range in countries with larger income. The reason for which is considered to be large diversity of lifestyles and wastewater treatment facility conditions especially when income is small. During the economic development process, PDC increases with pollutant generation per capita (PGC) increase with lifestyle change, however, PDC decreases after the development to some extent because of measures to reduce pollutant discharge and to improve water quality.

4.1 Chronological Alterations of Income, Pollutant Discharge and Water Quality

In developed countries, pollutant discharges have increased with economic development along with urbanization and industrialization (Tsuzuki 2006). The pollutant discharges sometimes exceed the ambient water environment capacity, which leads to water pollution problems with water quality deterioration. Some measurements would be conducted to reduce pollutant discharges in order to improve water quality.

Y. Tsuzuki, *Pollutant Discharge and Water Quality in Urbanisation*,
SpringerBriefs in Water Science and Technology,
DOI: 10.1007/978-3-319-04756-0_4, © The Author(s) 2014

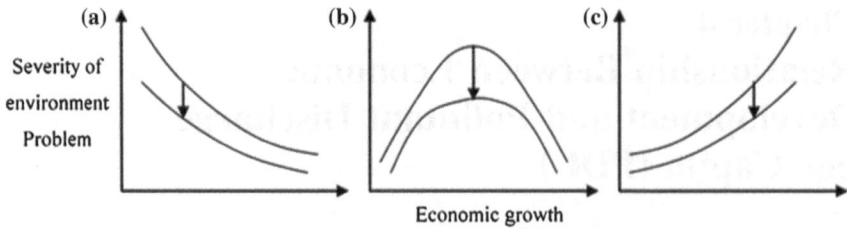

Fig. 4.1 Types of relationships between economic growth and severity of environment problems. Severity of environment problems can be downwarded in developing countries taking account of the experiences in developed countries. (Modified from Tsuzuki 2006, 2007; Original sources are Bai and Imura 2000, and World Bank 1992) (Copyright permissions have been obtained from Elsevier, JSCE and Nova Science Publishers) **a** Poverty-related issues (e.g. ratio of people lacking access to safe drinking water and appropriate sanitation). **b** Industrial pollution-related issues (e.g. particulate matter and SO_2 in air). **c** Consumption related issues (e.g. CO_2 discharge solid wastes generation amounts)

According to the World Bank (1992), the trends or changing petterns of economic development and severity of environmental problems depend on the type of environmental quality characteristics (Fig. 4.1) such as:

(a) poverty-related issues (severity of environmental problems decrease with economic development),
(b) industrial pollution-related issues (severity of environmental problems firstly increase with economic development, and secondly decrease with economic development). Relationships typically illustrated as Fig. 4.1b are called as the EKC.
(c) consumption-related issues (severity of environmental problems increase with economic development).

It should be worthwhile to investigate the relationships between parameters related to environmental problems and social and economical parameters to find solutions of the environment problems and academic purposes. In developing countries, it may be possible to avoid or limit the pollutant discharge increases after wise consideration of the experiences of developed countries. The economically developed countries have experienced several kinds of environmental problems along with economic development.

4.2 Relationship Between Economic Development and PDC

4.2.1 Relationship Between Economic Development and PDC in Japan

Chronological alterations of several PDC parameters have been summarized with income parameters in the drainage areas of Lakes Shinji and Nakaumi (Fig. 4.2)

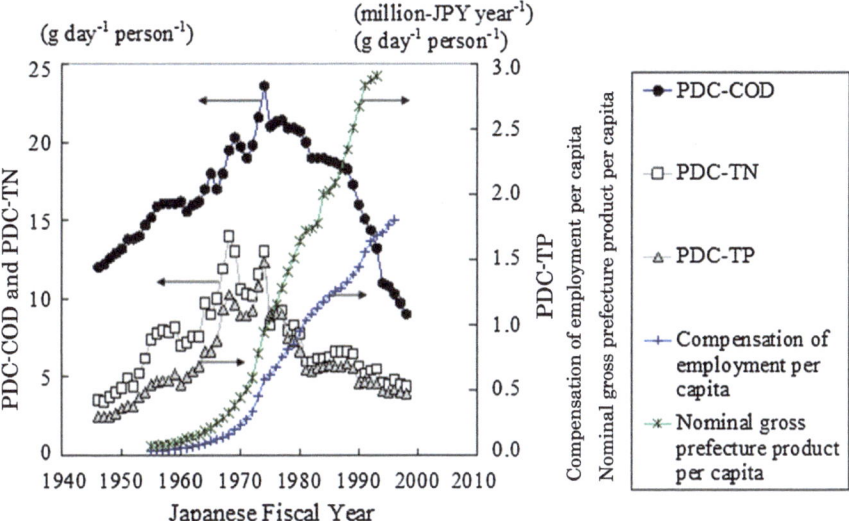

Fig. 4.2 Chronological alterations of pollutants discharge per capita (PDCs) and income indicators in river basin of Lakes Shinji and Nakaumi (Tsuzuki 2007, 2009a) (Copyright permissions have been obtained from JSCE and Nova Science Publishers)

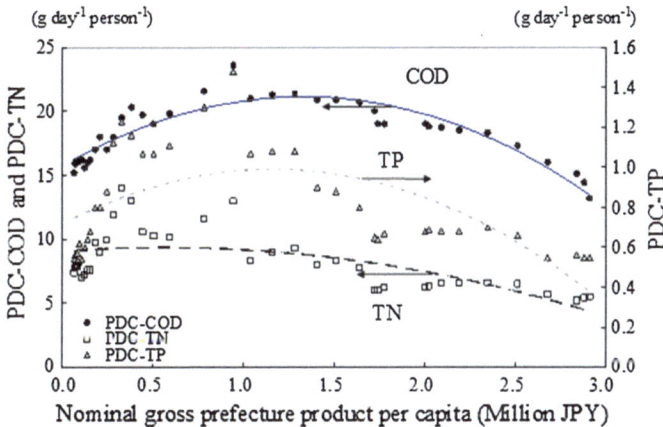

Fig. 4.3 Square-regression analyses of PDCs and gross prefectural production per capita in the river basin of Lakes Shinji and Nakaumi in 1955–93 (Tsuzuki 2007, 2009a) (Copyright permissions have been obtained from JSCE and Nova Science Publishers)

(Tsuzuki 2007, 2009a). PDC-COD$_{Mn}$, PDC-TN and PDC-TP have increased with the economic development in the 1940–1970s. Water pollution has been one of the most important environmental problems in Japan after the 1960s. A series of national environmental regulations including those of air and water environment

Table 4.1 Economics related indicators and PDC-BOD in the countries/areas in the coastal area (Tsuzuki 2007, 2009a) (Copyright permissions have been obtained from JSCE and Nova Science Publishers)

Region	Country	PPP-GNI[a] (U.S.$ 1,000) 2000	Access to sanitary facilities, household[b]		Access to safe drinking water, household[b]		Water usage for households per capita[c] (m³ person⁻¹ year) 1998-2002	WSEI[d]	Pollutant discharge per capita[e] (kg yr⁻¹ person⁻¹) BOD
			Total (%) 2002	Urban (%) 2002	Total (%) 2002	Urban (%) 2002			
		(a)	(b)	(c)	(d)	(e)	(f)	(g)	(h)
Pacific Islands	Papua New Guinea	2.3	45	67	11	61	7.2	0.04	1.6
South China Sea	Cambodia	1.4	16	53	6	31	4.3	0.01	18.2
	China	3.9	44	69	59	91	31.8	2.22	16.4
	Indonesia	2.8	52	71	17	31	30.5	0.12	14.8
	Malaysia	8.4	96	94	95	96	63.4	5.36	13.1
	Philippines	4.2	73	81	44	60	60.2	1.13	11.9
	Thailand	6.3	99	97	34	80	34.9	0.62	15.9
	Viet Nam	2.0	41	84	14	51	69.0	0.29	18.2
PORME Sea Region	Iran	5.9	84	86	87	96	72.9	4.97	0.5
	Kuwait	9.3	100	100	100	100	81.9	7.59	0.3
	Saudi Arabia	11.1	86	100	89	97	72.3	8.02	1.6
West and Central African (WACAF) Region	Northern WACFA	1.6	35	53	27	46	15.5	0.17	3.6
	Middle WACFA	1.0	40	53	14	27	13.5	0.02	1.7
	Southren WACFA	0.7	29	44	30	53	3.9	0.03	0.6
	Azerbadjan	2.8	55	73	47	76	100.0	2.46	11.5
	Iran	5.9	84	86	87	96	72.9	4.97	10.8

(continued)

Table 4.1 (continued)

Region	Country	PPP-GNI[a] (U.S.$ 1,000) 2000	Access to sanitary facilities, household[b]		Access to safe drinking water, household[b]		Water usage for households per capita[c] (m^3 person^{-1} year) 1998-2002	WSEI[d]	Pollutant discharge per capita[e] (kg yr^{-1} person^{-1}) BOD
			Total (%) 2002	Urban (%) 2002	Total (%) 2002	Urban (%) 2002			
Caspian Sea	Kazakhstan	5.5	72	87	61	88	38.2	1.80	1.1
	Turkmenistan	4.0	62	77	52	81	87.4	3.12	4.0
	Mombasa, Kenya	1.0	48	56	29	56	14.9	0.09	9.9
	Tanga, Tanzania	0.5	46	54	16	44	14.5	0.02	12.2
Eastern African Region	Dar es Salaam, Tanzan	0.5	46	54	16	44	14.5	0.02	4.4
	Seychelles	7.3	100	100	87	100	2.0	0.13	19.5
	Madagascar	0.8	33	49	5	14	24.9	0.01	2.2
	Comoros	1.5	23	38	25	47	6.4	0.13	6.0
Red Sea and Gulf of Aden	Jeddah, Saudi Alabia	11.1	86	100	89	47	72.3	3.89	1.3

[a] World Bank (1992)

[b] World Resource Institute (2006) Earth Trend Environmental Information (available on http://earthtrends.wri.org/searchable_db, accessed in April, 2006) Source World Health Organization (WHO) and United Nations Children's Fund (UNICEF) 2004. WHO/UNICEF Joint Monitoring Programme for Water Supply and Sanitation; Meeting the MDG drinking water and sanitation target: a mid-term assessment of progress

[c] Estimated from population and domestic water use in Food and Agriculture Organization of the United Nations (FAO) Land and Water Development Division. 2005. AQUASTAT Information System on Water and Agriculture: Online database. Rome: FAO. Available on-line at http://www.fao.org/ag/agl/aglw/aquastat/dbase/index. stm

[d] Water, sanitation and economy index defined by equation (1); and

[e] Calculated by authors based on the reports from UNEP (Tsuzuki 2004)

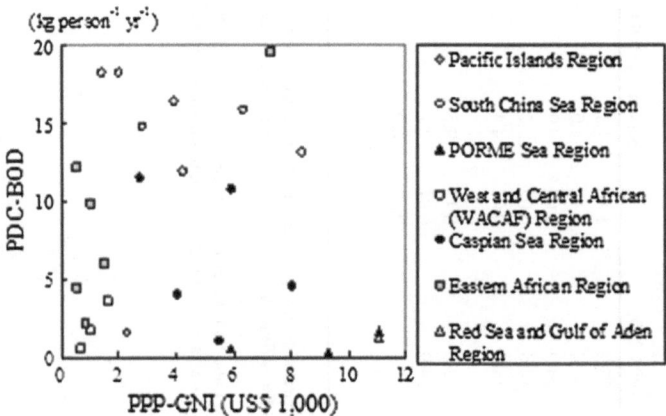

Fig. 4.4 Purchase power parity based gross national income (PPP-GNI) per capita and PDC-BOD in countries/areas in coastal areas (Tsuzuki 2007, 2009a) (Copyright permissions have been obtained from Elsevier and Nova Science Publishers)

Fig. 4.5 Relationship between PPP-GNI and PDC-BOD in countries/areas in the coastal areas: the Pacific Island, South China Sea, and African Regions (Tsuzuki 2008b) (Copyright permissions have been obtained from Elsevier and Nova Science Publishers)

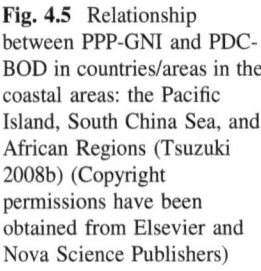

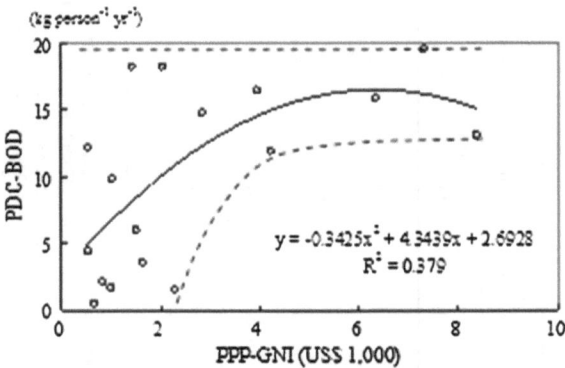

has been established in 1970. After that, environmental pollution has been alleviated based on these regulations and their implementations.

The chronological relationships between PDCs and income indicators have been found to be the EKC's inverted U-shaped curves (Fig. 4.3) (Tsuzuki 2007, 2009a). Larger correlation of PDC-COD$_{Mn}$ with the income indicator than PDC-TN and PDC-TP has suggested that the economic development has been a strong factor to determine PDC-COD$_{Mn}$ in Japan. PDC-TN and PDC-TP discharge reductions should have needed more improvement in wastewater treatment technology and economic development level. One of the reasons of the differences between organic carbons and nutrients is that wastewater treatment has been struggled and achieved in developed countries along with technological developments. Developing countries can adopt and use these technologies under the specific social and economic conditions and sustainability of these technologies in the countries.

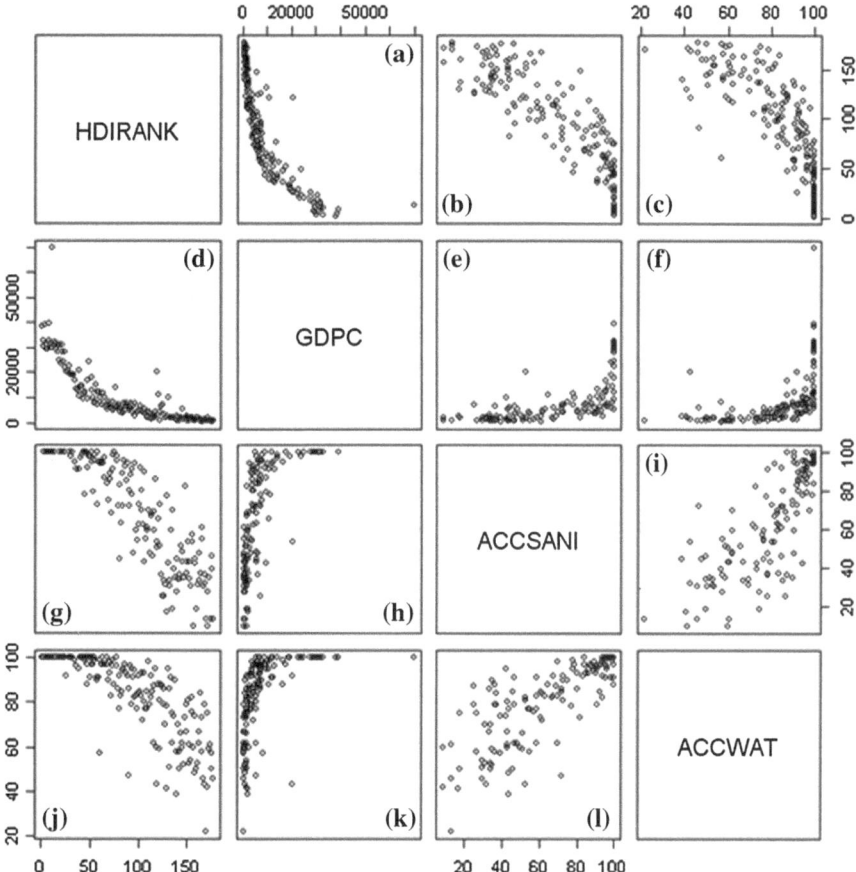

Fig. 4.6 Relationships between Human Development Indicator Rank (HDIRANK), GDP per capita (GDPC), proportion of population with access to appropriate sanitation (ACCSANI) and safe water (ACCWAT) in 2004 (Prepared by the author using UNDP data: UNDP 2007) (Tsuzuki 2008b) (Copyright permissions have been obtained from Elsevier and Nova Science Publishers)

4.2.2   Relationship Between Economic Development and PDC in Developing Countries

When relationships between municipal wastewater pollutant discharges and economic development indicators are considered, the MDGs water and sanitation indicators (accesses to safe drinking water and appropriate sanitation) and household water usage amount are anticipated to influence PDCs. A simple relationship between PDC-BOD and purchase power parity based gross national income (PPP-GNI) per capita has not been found for countries/areas in the Pacific Islands, Southeast Asia, Middle East, Caspian Sea and Africa Regions (Table 4.1, Fig. 4.4) (Tsuzuki 2007).

However, only for the countries/areas in Asia, Pacific and Africa, where the socio-economical and environmental conditions are considered to be similar and within certain ranges, the relationship has been found to be inverted U-shaped curve (Fig. 4.5). The relationship has been similar to the original Kuznets relationship between income and inequity (Kuznets 1955), i.e. income inequity is large when income is small, and income inequity is small when income is large.

In some developing countries, the MDGs water and sanitation indicators are still struggling targets especially for the sanitation indicators, which is sometimes regarded as hard to be achieved in 2015. Water and sanitation is directly linked to poverty and starvation. The relationships of the MDGs indicators with the income indicator and the Human Development Indicator (HDI) Rank are shown in Fig. 4.6 (Tsuzuki 2008b). Economic development level expressed with GDP per capita (GDPC) and life style improvement expressed with HDI are both related to the MDGs water and sanitation indicators (Fig. 4.6b, c, e, f, g, h, j and k). For water and sanitation development, the MDGs water and sanitation indicators should be focused and attained by 2015. When wastewater treatment beyond 'sanitation' is considered, quantitative analysis should be required to illustrate water quality and pollutant loads as well as such indicators as PDC (Chap. 2).

References

Bai X, Imura H (2000) A Comparative study of urban environment in East Asia: stage model of urban environmental evolution. Int Rev Environ Strat 1(1):135–158

Grossman GM, Krueger AB (1995) Economic growth and development. Q J Econ 110(2):353–377

Kuznets S (1955) Economic growth and income inequality. Am Econ Rev 45(1):1–28

Tsuzuki Y (2004) Land based water pollutant loads of Tokyo Bay and nutrients removal by fishery, environment system. In: Proceedings of 32nd annual meeting of environmental systems research. pp. 413-418 (in Japanese with English abstract)

Tsuzuki Y (2006) An index directly indicates land-based pollutant load contributions of domestic wastewater to the water pollution and its application. Sci Total Environ 370(2–3):425–440

Tsuzuki Y (2007) Relationships between pollutant discharges per capita (PDC) of domestic wastewater and the economic development indicators. J Environ Syst Eng Jpn Soc Civ Eng 63(4):224–232 (in Japanese with English abstract)

Tsuzuki Y (2008b) Relationships between water pollutant discharges per capita (PDCs) and indicators of economic level, water supply and sanitation in developing countries. Ecol Econ 68:273–287

Tsuzuki Y (2009a) Relationships between pollutant discharges per capita (PDC) of domestic wastewater and the economic development indicators. J Environ Syst Eng Jpn Soc Civ Eng 14:37–46

Tsuzuki Y (2009c) Comparison of pollutant discharges per capita (PDC) and its relationships with economic development: an indicator for ambient water quality improvement as well as the Millennium Development Goals (MDGs) sanitation indicator. Ecol Ind 9:971–981

UNDP (2007) Statistics of the Human Development Report. http://hdr.undp.org/en/statistics/

World Bank (1992) World development report 1992. Oxford University Press, New York pp 308

Chapter 5
Pollutant Discharge Control of Municipal Wastewater

In this chapter, municipal wastewater pollutant discharge control methods in Japan are explained and compared to those in Thailand (Tsuzuki et al. 2009a; Tsuzuki 2011). The Millennium Development Goals (MDGs) sanitation indicator, proportion of people with access to appropriate sanitation in Thailand have been already 99–100 %, however, PDCs in Thailand is larger than those in Japan. More improvement of wastewater treatment is necessary even in developing countries especially those with large MDG sanitation indicator such as Thailand. A part of this chapter is based on Tsuzuki (2011).

5.1 Municipal Wastewater Pollutant Discharge Control in the Urban and Peri-Urban Areas: A Comparison Between Thailand and Japan

Current municipal wastewater treatment schemes in Japan are characterized with (1) centralized WWTPs in urban areas, and (2) on-site WWTPs in rural areas. On-site WWTPs are also applied in low-population density areas in urban areas (Gaulke 2006; Tsuzuki 2006). New installation of on-site WWTPs is limited by regulations to only combined *johkasou* (*jokaso*) (CJ) (*gappei-shori johkasou* in Japanese). The regulations have been enforced for a decade. Some old type on-site treatment systems are still in use: simple *johkasou* (SJ) (*tandoku-shori johkasou*) and night soil treatment (NST) (*kumitori-benjo*). SJ and NST treat only black water.[1] Grey water[2] is discharged without treatment in SJ and NST. NST is similar to septic tanks which are often applied in developing countries.

Sedimentation or septage in the NST underground tank in Japan is collected with a vacuum car or other relevant methods, and the collected septage is treated at NST wastewater and SJ sludge treatment plants or WWTPs, or utilised as agriculture fertiliser. Energy recycles with methane fermentation has been conducted

---

[1] Toilet wastewater.
[2] Wastewater from kitchen, bath, washing clothes and so on.

Y. Tsuzuki, *Pollutant Discharge and Water Quality in Urbanisation*,
SpringerBriefs in Water Science and Technology,
DOI: 10.1007/978-3-319-04756-0_5, © The Author(s) 2014

Table 5.1 Pollutant discharge derived from municipal wastewater in Chiba City, Japan

Pollutant source and treatment method	Pollutant generation per capita (PGC) and pollutant discharge per capita (PDC) (g person^{-1} day^{-1})			
	BOD	COD	TN	TP
PGC or basic units of domestic wastewater	45	23	8.5	1.0
Night soil (blackwater)	16	10	7.0	0.7
Kitchen, bath and washing clothes (greywater)	29	13	1.5	0.3
PDC by domestic wastewater treatment methods				
Combined *jokaso* (CJ)	3.2	4.6	7.0	0.88
Simple *jokaso* (SJ)	32.2	16.5	7.5	0.97
SJ, derived from blackwater	3.2	3.5	6.0	0.67
SJ, derived from greywater	29.0	13.0	1.5	0.30
Night soil treatment (NST)	29.3	14.0	4.3	0.52
NST, derived from treatment plant effluent	0.3	1.0	2.8	0.22
NST, derived from greywater	29.0	13.0	1.5	0.30

(Tsuzuki 2008a, 2011; Calculated by the author based on Fujimoto 1988, Fujimura 1996, and Fujimura and Nakajima 1998) (Copyright permissions have been obtained from Elsevier and Nova Science Publishers)

in a few advanced NST wastewater and SJ sludge treatment plants. PGCs and PDCs of on-site municipal wastewater treatment systems in Japan have been summarized in Table 5.1 (Tsuzuki 2008a, 2011).

Comparison of PDC in developing and developed countries, Thailand and Japan, shows larger PDC in Thailand than Japan (Fig. 5.1) (Tsuzuki et al. 2008a, b). The MDG sanitation indicator in Thailand is already 99–100 %. However, PDCs in suburban area of Bangkok has been estimated to be larger than those in Japan especially for PDC-BOD (Fig. 5.1), which means that more improvement of municipal wastewater treatment systems may be necessary to improve the ambient water quality in Bangkok, Thailand.

Total pollutant discharges and PDC of the Pack Kret Municipality, sub-urban of Bangkok, have been estimated based on the material flux (flow) analysis (MFA) results (Sinsupan and Koottatep 2004). BOD discharge per capita (PDC-BOD) has been estimated as 48.4 gpd when *seepage and septage of septic tank and leachate of composting* (SSL) is included, and 29.2 gpd without SSL. These estimated PDC can compare with the exisiting PDC estimation in Thailand based on the UNEP report, 43 gpd (Tsuzuki 2006), and pollutant generation per capita (PGC) applied in World Bank (2001), 35 gpd. The PDC estimation results reveal that relatively large amounts of PDC in SSL should contribute much in total pollutant discharges from municipal wastewater (Fig. 5.2) (Tsuzuki et al. 2008a).

SSL are usually not considered so much in pollutant discharge and pollutant load estimation. However, SSL can infiltrate into soil and groundwater and consistently cause and contribute to ambient water quality deterioration. Moreover, groundwater contaminations of the pollutants do harm to groundwater, which will cause water supply problems if groundwater is used as water supply sources for households and industries.

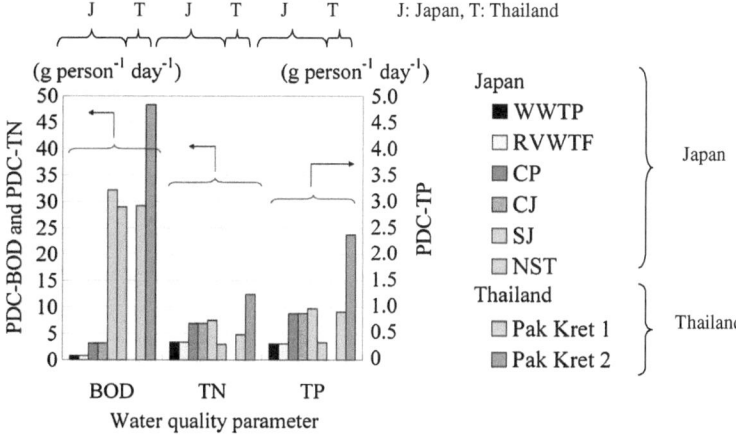

Fig. 5.1 Pollutant discharges per capita (PDCs) in Japan estimated based on the existing treatment facilities in Chiba City, Japan and those in Pak Kret Municipality, sub-urban of Bangkok, Thailand (Tsuzuki et al. 2008a, b. http://www.wsscc.org/resources/resource-pub lications/water-quality-and-pollutant-load-ambient-water-and-domestic) (Copyright permissions have been obtained from JSCE and JECES)

5.2   Centralised and On-site Municipal Wastewater Treatment in Japan

After the World War II, Japan has experienced rapid economic development. Sanitation system has also developed during the period. Comparing to other developed and developing countries, Japan currently has a kind of a unique and independent sanitation systems which consists of centralized wastewater treatment plants (WWTPs) and onsite treatment systems called combined *johkasou*.[3] The planning areas of centralised and on-site WWTPs are divided in Japan. Newly installations of municipal wastewater treatment facilities have been recently limited by regulations only to centralized WWTPs and combined *johkasou*, however, there are still not so small amount of old type sanitation system stock, namely simple *johkasou* and night soil treatment systems.

---

[3] *Johkasou* is a common name of on-site treatment system in Japan and also used as a trade name of a company product (http://www.gec.jp/WATER/, http://www.wepa-db.net/technologies/individual/list_b_onsite.htm, accessed on 13 April 2010). There are two types of *johkasou*, combined *johkasou* and simple *johkasou*. Combined *johkasou* treats both black and gray water, and simple *johkasou* treats only black water. Only a word, *johkasou*, currently means combined *johkasou*. *Johkaso* and *jokaso* are also sometimes used.

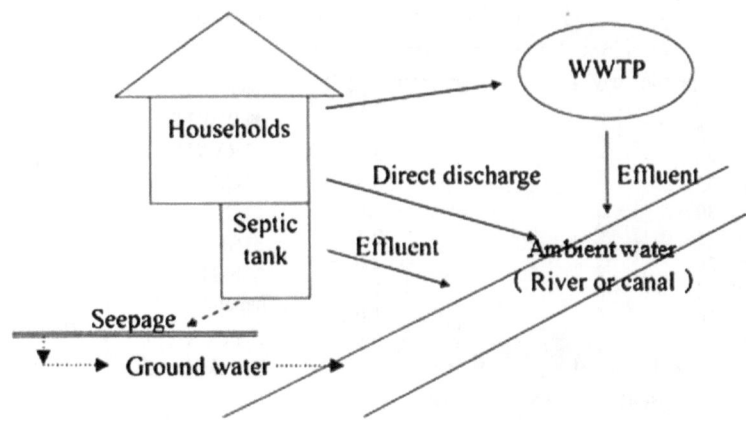

Fig. 5.2 PDCs estimation scheme in Pak Kret Municipality, Thailand (Modified from Tsuzuki et al. 2008a) (Copyright permissions have been obtained from JECES)

Besides there have been domestic conflicts and discussion for the basin wide WWTPs in the 1980s including economical feasibility of centralized and on-site municipal wastewater systems and river water deficiency between a water intake point of water supply plants and a water discharge point of WWTPs (Nakanishi and Okinoto 1982), zoning or geographical dividing concept of centralized and on-site municipal wastewater treatment systems have been established. Wastewater treatment system zoning is mainly based on population density and specific characteristics of areas including agriculture and fishery communities. The zoning is generally conducted at prefecture[4] scale. Centralized WWTPs with piped-connection systems to collect municipal and some commercial and industrial wastewater are developed in urban areas, and on-site treatment systems called combined *johkasou* are constructed in less-habitat or rural areas (Infrastructure Development Institute, Japan 2004; Tsuzuki 2009b). The population density based scheme of centralized and on-site treatment zoning is explained in the official guidelines of the Ministry of Land, Infrastructure, Transportation and Tourism (Basin Wide Sewerage Comprehensive Planning System Design Council 2008).

The actual deployment planning of municipal wastewater treatment at prefectural scale is consisted of the areas for basin wide WWTPs, local government WWTPs, specific environment protection public WWTPs, fishery community WWTPs, agriculture community WWTPs, community plants and combined *johkasou*. Community plants are something like large scale on-site treatment systems. Regulations of these centralized WWTPs are managed by each corresponding central government ministries. There have been 166 basin wide WWTPs, 1200 local government WWTPs and 736 specific environment protection public WWTPs at the end of March, 2009 (Ministry of Land, Infrastructure,

---

[4] There are 47 prefectures in Japan.

Table 5.2 People with advanced municipal wastewater treatment in Japan in 2003–05

Advanced wastewater treatment method	Population served with advanced municipal wastewater treatment (1,000 persons)		
	March 2003	March 2004	March 2005
Public WWTPs[a]	82,570	84,580	86,360
Agricultural community WWTPs[b]	3,110	3,280	3,440
Combined *johkasou*	9,930	10,300	10,620
Community plants	380	380	370
Total[c]	95,990	98,540	100,790
Proportion of people served with advanced municipal wastewater treatment (%)	75.8 %	77.7 %	79.4 %
Total population[d]	126,690	126,820	126,870

(Tsuzuki 2011) (Copyright permissions has been obtained from Nova Science Publishers)
[a] Basin wide WWTPs and local government WWTPs
[b] Agriculture community WWTPs, fishery community WWTPs, forest community WWTPs, and specific environment protection public WWTPs
[c] Total population with advanced municipal wastewater treatment
[d] Total population in Japan
Prepared by the author based on Ministry of the Environments (2005)

Transportation and Tourism, Japan 2009). These three kinds of WWTPs are generally categorized as public WWTPs. For agriculture community WWTPs, development and construction has been completed at 5,060 communities among 5,400 communities with their planning and design in 2007 (Ministry of Agriculture, Forestry and Fisheries 2008b).

The areas of several kinds of WWTPs are zoned or geographically divided depending on population density and geographical characteristics. For example, population with "appropriate" sanitation has been 1,216 thousand among 2,374 thousand total population in fishery villages in 2009 (Ministry of Agriculture, Forestry and Fisheries 2009a). The definition of "appropriate" sanitation in Japan is not the same as "appropriate" sanitation of the world sanitation perspectives. Therefore, "advanced" sanitation is used to describe "appropriate" sanitation of the current Japanese sanitation perspectives in this book. Simple *johkasou* and night soil treatment systems, which treat only black water,[5] are not included in "advanced" sanitation in Japan (Tsuzuki 2006). Gray water[6] is directly discharged to ambient water without treatment by simple *johkasou* and night soil treatment systems. Among the population with advanced sanitation in fishery villages of 1,215 thousands, 641 thousands with public WWTPs (53 %), 319 thousands with combined *johkasou* (26 %), 173 thousands with fishery community WWTPs (14 %), and 82 thousands with agriculture community WWTPs (7 %) at the end of

---

[5] Toilet wastewater.

[6] Wastewater from households other than toilet including kitchen, bathing and washing clothes and so on.

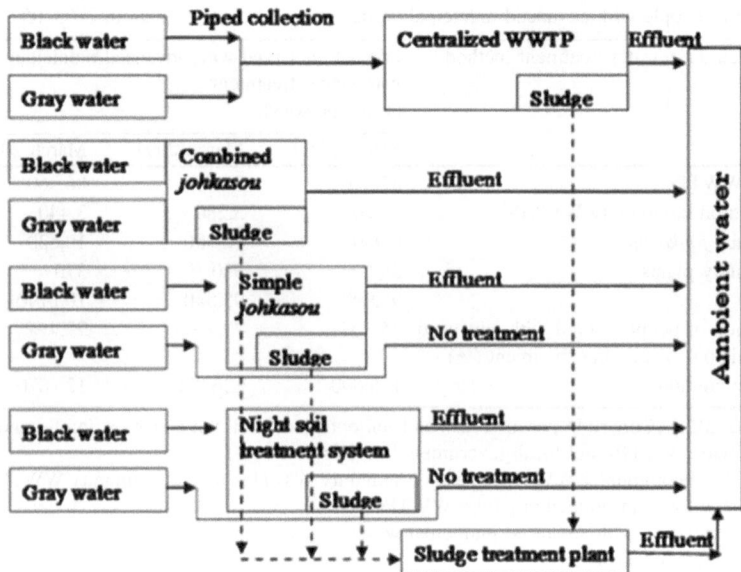

Fig. 5.3 Flow of black and gray water with municipal wastewater treatment in Japan (Tsuzuki 2011) (Copyright permissions has been obtained from Nova Science Publishers)

March, 2009. These numbers show that the geographical planning or zoning of municipal wastewater treatment is designed in detail and fishery villages are consisted of miscellaneous areas in regards to population density and geographical characteristics. National average of percentage of population with advanced sanitation has been 85 % and that in fishery villages has been 51 % in March, 2009. Populations with advanced municipal wastewater treatment including public WWTPs, agriculture community WWTPs and combined *johkasou* have been increasing every year (Table 5.2). From the perspective of the MDGs sanitation indicator, people with appropriate sanitation is already 100 % in Japan. The MDGs indicator figure includes both simple *johkasou* and night soil treatment systems.

Figure 5.3 shows flow of black and gray water and sludge by use of centralized WWTPs, combined *johkasou*, simple *johkasou*, and night soil treatment systems. Among three on-site wastewater treatment methods, only combined *johkasou* can be newly constructed by the current regulations in Japan. However, simple *johkasou* and night soil treatment systems which treat only black water are still working and re-construction of these old type on-site treatment systems into combined *johkasou* is a challenging topic in rural area (Tsuzuki 2006). In urban area with existing centralized WWTPs, piped connection development is a challenging topic, which changes on-site treatment systems to centralized WWTPs.

Construction and maintenance costs of centralized municipal wastewater treatment systems including WWTPs and piped collection systems are covered by wastewater treatment tariff which is usually calculated based on water supply

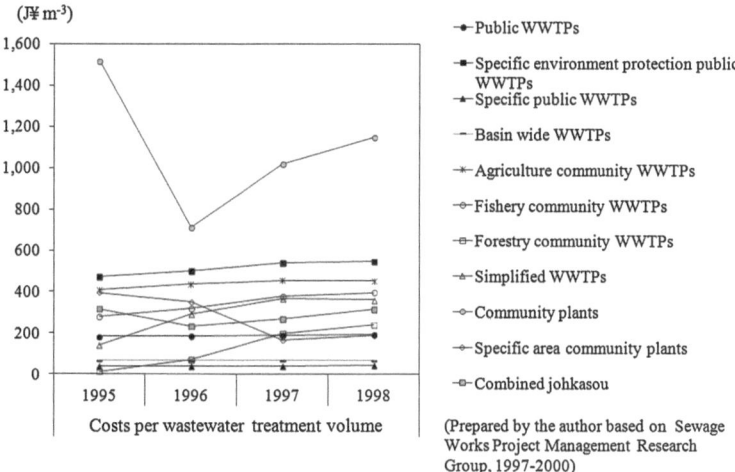

(J¥ m⁻³)

- ◆ Public WWTPs
- ■ Specific environment protection public WWTPs
- ▲ Specific public WWTPs
- − Basin wide WWTPs
- ✳ Agriculture community WWTPs
- ○ Fishery community WWTPs
- ⊟ Forestry community WWTPs
- ▲ Simplified WWTPs
- ◇ Community plants
- ● Specific area community plants
- ⊟ Combined johkasou

Costs per wastewater treatment volume

(Prepared by the author based on Sewage Works Project Management Research Group, 1997-2000)

Fig. 5.4 Construction and maintenance costs per wastewater treatment volume in Japanese Fiscal Year 1995–1998 (J¥ m⁻³) (Tsuzuki 2011) (Copyright permissions has been obtained from Nova Science Publishers)

usage amounts and government subsidies. For on-site treatment systems, combined *johkasou*, construction costs are covered by house owners and subsidies by governments. Subsidies are provided by governments for on-site treatment systems because of financial efficiency. On-site treatment systems are cost effective comparing to centralized wastewater treatment systems with piped collection systems in rural areas with low population density. Maintenance cost of on-site treatment systems including electricity cost for aeration blower operation and payment for de-sludging are usually covered by house owners. Construction and maintenance costs per wastewater treatment volume of several kinds of municipal wastewater treatment systems in Japanese Fiscal Year 1995–98 have been J¥ 15–1516 m⁻³, and J¥ 43–1148 m⁻³ in 1998 (Fig. 5.4) (Sewerage Works Project Management Research Group, 1997–2000). There have also been differences in costs per wastewater treatment volume even in the same treatment methods. Therefore, benefit-cost analyses are conducted for each WWTP project (Table 5.2).

5.3 Pollutant Removal Efficiencies of Centralised Wastewater Treatment Plants

For centralised wastewater treatment plants (WWTPs), there are a lot of research on how effluent water quality is determined and how it can be improved. Most treated wastewater is subject to the water quality standards or criteria for the WWTP effluent, which should be complied by regulations. Centralised WWTPs is a huge system including wastewater collection piping systems, WWTPs, and sludge treatment systems. Subjected areas are sometimes under development and WWTP system development should be accordance with the land, housing and industry development. When these developments and WWTP system development are not correspondent, WWTP systems are sometimes excess or lack of capacity. Such mismatches sometimes lead to lower treatment efficiency at WWTPs including smaller pollutant removal efficiencies, and larger concentration of WWTP effluent. These problems should be avoided in the stage of basic designing.

Biological processes are usually applied in the centralised WWTPs including activated sludge process, aerated lagoon, stabilisation reservoir and pond, and rotating biological contactor (RBC) (Tsuzuki 2012). In most biological treatment systems such as activated sludge process and ecological treatment systems, effluent pollutant concentration is often expressed as Eq. 5.1. Therefore, removal efficiency of a pollutant at centralised WWTPs can be expressed as a function of reciprocal of pollutant influent concentration (Eq. 5.2).

$$C_{ef} = A \times C_{in} + B \tag{5.1}$$

where C_{ef} is effluent pollutant concentration/load (g m^{-3} or g day^{-1}); C_{in} is influent pollutant concentration/load (g m^{-3} or g day^{-1}); a is a coefficient (unitless); and b is a coefficient (g m^{-3} or g day^{-1}).

Removal Efficiency (%) = $(1 - \frac{C_{ef}}{C_{in}}) \times 100$

$$= (1 - A - \frac{B}{C_{in}}) \times 100 \tag{5.2}$$

When A equals 0.1, relationship between pollutant removal efficiency and influent pollutant concentration can be shown as Fig. 5.5. When A equals to 0.1, maximum pollutant removal efficiency is 90 % when B equals to zero. When B equals from 20 to 100, pollutant removal efficiencies increase with the increase of pollutant. It means influent pollutant concentrations should be kept larger than a certain concentration. If the influent pollutant concentration is much less than a certain concentration, pollutant removal efficiency decreases. Under the mismatch conditions of the development of pollutant sources and WWTP system development, a pollutant effluent concentration may be smaller than the effluent standards or criteria, however, the pollutant removal efficiency is small and the WWTP system is totally low efficiency. Centralised WWTP system is a huge system and costs much. Appropriate designing especially on the WWTP treatment capacity is a critical and important.

Fig. 5.5 Relationship between influent pollutant concentration and removal efficiency with Eq. 5.2. (A is assumed to be 0.1.) (Modified from Tsuzuki 2012) (Copyright permission has been obtained from IWA)

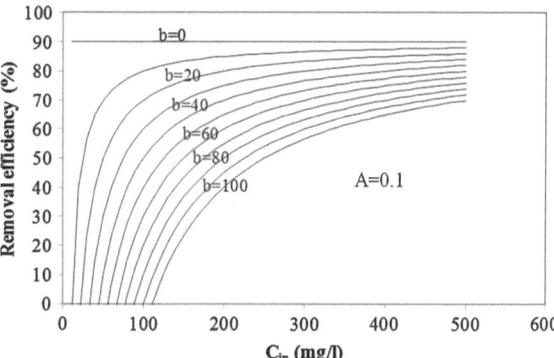

5.4 Areal Deployment Strategies of Centralized and On-site WWTPs: Zoning or Intermixed

Areal deployment of sanitation facilities in Japan is zoned or geographically organized at each prefectural level as described above, and some basin wide WWTPs include areas of more than one prefecture (Infrastructure Development Institute, Japan 2004; Tsuzuki et al. 2009b). On the contrary, in Thailand, middle-class developing countries with the MDGs sanitation indicator of 96 % in 2006, on-site treatment systems such as septic tanks and cesspools are commonly applied in both urban and rural areas, even in the areas served with centralized WWTPs (Tsuzuki et al. 2009b). Besides large value of the MDG sanitation indicator, ambient water pollution is still a major environment problem in Thailand, especially highly populated central area including the capital, Bangkok. Based on the results of pollutant discharge estimation of municipal wastewater, pollutant discharge per capita (PDC) in Pak Kret District, sub-urban of Bangkok, has been estimated as 48.4 g-BOD person^{-1} day^{-1} with seepage and septage of septic tank and leachate from composting (SSL) and 29.2 g-BOD person^{-1} day^{-1} without SSL based on the material flow analysis (MFA) results (Tsuzuki et al. 2008b). These PDC values have been larger compared to those in Japan and found to be a reason for ambient water deterioration.

Systematic and strategic planning scheme of municipal wastewater treatment should be prepared for the mixture condition of on-site treatment systems and centralized WWTPs. WWTPs influent sometimes include certain industrial wastewater. The intermixed condition is not commonly found in developed countries, however, may not be so unique in developing countries where on-site WWTPs development has been preceded to centralized WWTPs development. Simple on-site WWTPs are required to prevent infectious diseases and to improve people's lifestyles when there is no centralized WWTP.

After the concept of step development from on-site treatment to centralized treatment in a draft guidelines of sanitation development in developing countries (Infrastructure Development Institute, Japan 2004), scenario-based approaches including estimation of PDC have been proposed for improvement of the mixture condition of on-site and centralized municipal wastewater treatment systems (Tsuzuki et al. 2009a; Suharyanto and Matsushita 2009). Pollutant discharge and pollutant load analyses including both on-site and centralized wastewater treatment systems in the project planning stage are important to improve ambient water quality as well as increasing percentage of people with appropriate sanitation.

5.5 Quantitative Evaluation of Pollutant Discharge from Municipal Wastewater Under the Mixture Conditions of Centralised and On-site Wastewater Treatment Systems: A Case Study in Bangkok, Thailand

In Japanese designing and planning scheme of municipal wastewater treatment systems (WWTSs), the areas of centralised WWTPs and those of on-site WWTPs are divided. Practically, some old on-site WWTPs are remaining in the centralised WWTP areas, and the houses with these on-site WWTPs are recommended to connect their wastewater treatment systems to piped collection systems of centralised WWTPs. In the areas of on-site treatment systems, governments sometimes subsidise to installation of advanced on-site treatment system, combined *johkasou*, or advanced combined *johkasou* (Tsuzuki 2011). These installations are usually upgrade of on-site WWTP from septic tank or old type of simple *johkasou*. Thus, municipal wastewater is usually treated at centralised WWTP or on-site WWTP and not treated with both on-site and centralised WWTPs in Japan. Detailed information on Japanese on-site WWTPs and WWTSs can be found in the existing literature (Gaulke 2006; Tsuzuki 2006).

On the contrary, in Bangkok, Thailand, even in the areas of centralised WWTPs, a septic tank or other types of on-site WWTPs is generally mandatory installed by regulations (Tsuzuki et al. 2009b). There are several centralised WWTPs in Bangkok areas and it is worth to consider whether the existing mixture systems should be continued or dividing systems in Japan should be introduced.

For the purposes of investigate this point, material flow analysis on pollutant discharge from municipal wastewater in the housing using on-site treatment systems has been conducted to estimate pollutant discharge amounts, and pollutant removal efficiency function has been developed (Tsuzuki et al. 2013a). Based on these results, scenario-based estimation of pollutant discharge from municipal wastewater have been conducted to estimate total pollutant discharge amounts with improvement of centralised and on-site WWTSs (Tsuzuki et al. 2013b).

Typical pollutant discharge amounts from municipal wastewater with on-site WWTP in peri-urban areas of Bangkok have been determined as shown in Fig. 5.6 (Tsuzuki et al. 2013a). Pollutant discharge from municipal wastewater using on-site WWTP is considered to be total of Effluent (PDC_{eff}), Septage (PDC_{sep}) and Seepage (PDC_a). Seepage (a) is pollutant discharge from on-site WWTP to

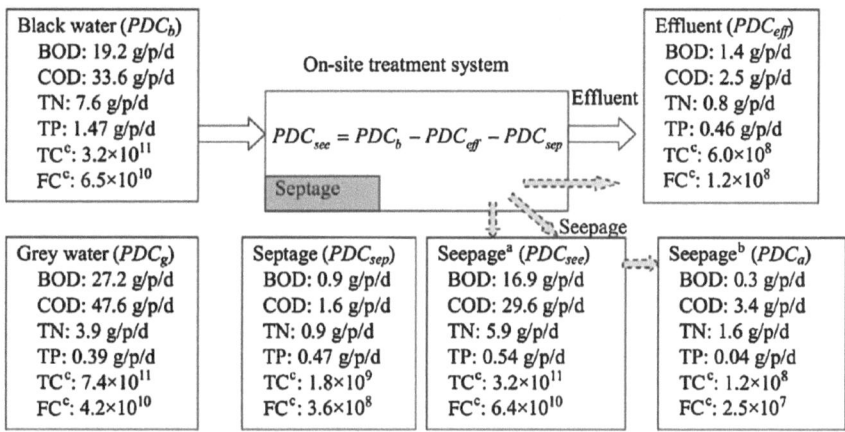

ᵃPDC of seepage, *PDC_see*, was estimated based on the following equation without consideration of degradation during storage in the on-site treatment system.

$$PDC_{see} = PDC_b - PDC_{eff} - PDC_{sep}$$

ᵇPDC to ambient water through soil and groundwater. Effects of degradation in on-site treatment system and soil were considered.

ᶜUnits of TC and FC are mean probable number (MPN) per person per day.

Fig. 5.6 Estimation results of pollutant discharge per capita (PDC) of on-site treatment systems in peri-urban areas of Bangkok, Thailand (*Source* Tsuzuki et al. (2013a). Estimated based on Sinsupan and Koottatep (2004), Jiawkok and Koottatep (2006), and Tsuzuki et al. (2009b)) (Copyright permission has been obtained from IWA)

seepage. Seepage (b) is pollutant discharge of seepage which is flown to ambient water including the rivers and canals through soil and groundwater. Pollutant removal efficiencies of pollutants at centralised WWTPs have been expressed with several equations (Fig. 5.7).

Based on these results, total pollutant discharge amounts have been investigated with scenario-based analysis method (Fig. 5.8) (Tsuzuki et al. 2013b). Scenario 1 is to develop both on-site and centralised WWTSs continuing the existing mixture conditions. Scenario 2 is to divide the centralised and on-site WWTS areas. Based on the conditions of on-site and centralised WWTSs, the conditions of canal water should also be considered when designing and planning of both on-site and centralised WWTSs are developed.

5.6 Technological Alternatives in Consideration with Pollutant Discharge and Investment

UNDP data in 2004 (UNDP 2007) show that the MDGs water and sanitation indicators are related to Human Development Indicator Rank and Gross Domestic Product (GDP) per capita (Tsuzuki 2008a). Economic development in a country is considered to be necessary to increase water and sanitation indicators, percentage

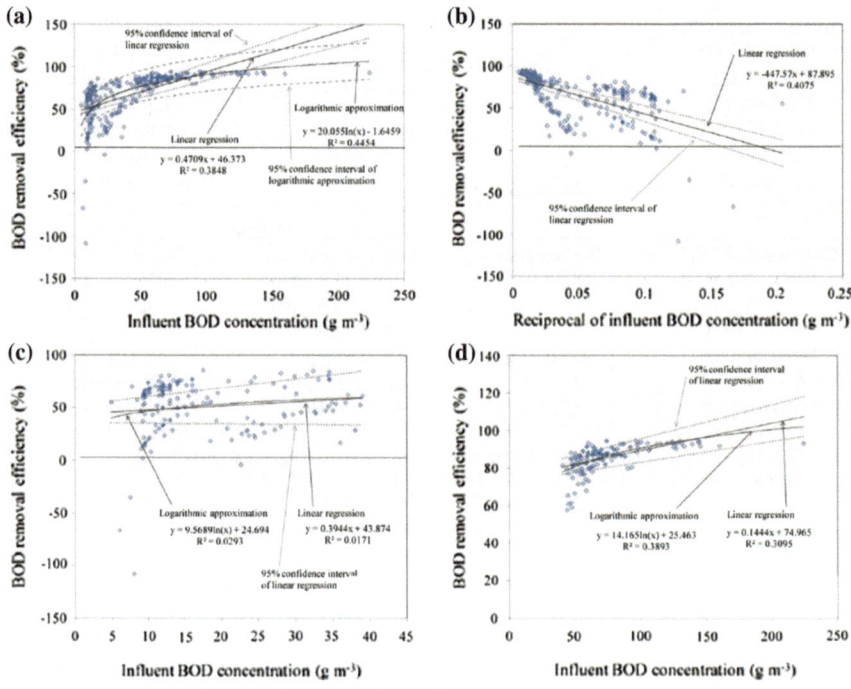

Fig. 5.7 BOD removal efficiency (%) as functions of (**a**) influent BOD concentration (**b**) reciprocal of influent BOD concentration (**c**) influent BOD concentration in a range of 0–40 g-BOD m^{-3}, and (**d**) influent BOD concentration in a range of >40 g-BOD m^{-3}. (Tsuzuki et al. 2013a) (Copyright permission has been obtained from IWA)

of people with access to safe drinking water and appropriate sanitation. There are many countries with GDP per capita of less than US\$ 10,000 person^{-1} year^{-1}. In these countries, the MDGs water and sanitation indicators are small. On the contrary, most countries with GDP per capita with more than US\$ 10,000 are with large values of the MDGs water and sanitation indicators. Therefore, US\$ 10,000 is considered to be a rough indicator to improve water and sanitation conditions at the MDGs indicator levels.

Low-cost is a key word of sanitation in developing countries. Technical alternatives have been summarized as described above (Franceys et al. 1992; Serageldin 1994; Mara 1996a, b; Infrastructure Development Institute, Japan 2004; Tilley et al. 2008). For *johkasou* in Japan, a concrete structured on-site treatment system with anaerobic and aerobic chambers has been developed around 1990 (Ishii and Yamada 1990). A lot of lactic acid beverage plastic bottles of about 20 ml each have been used for biofilm attachment structures in the chambers to form biofilm on their surface and to enhance microorganism activities. The effluent wastewater quality of the so-called Ishii-type *johkasou* has been as clean as drinkable. However, that type of an on-site treatment system has not been admitted by the government inspection because only structure based standards have been

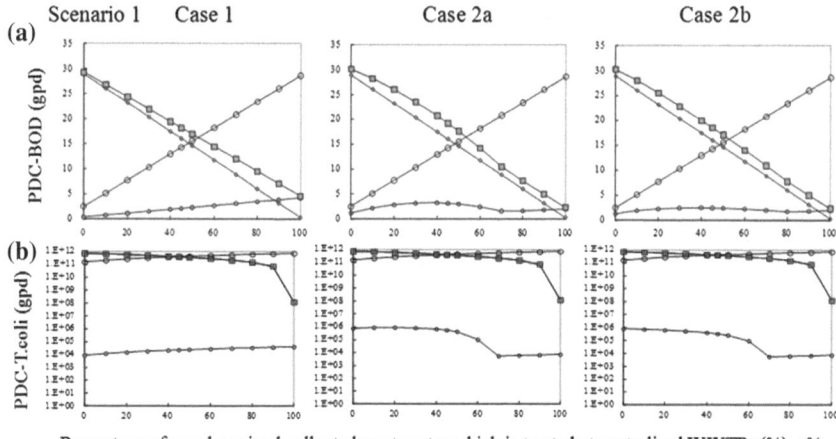

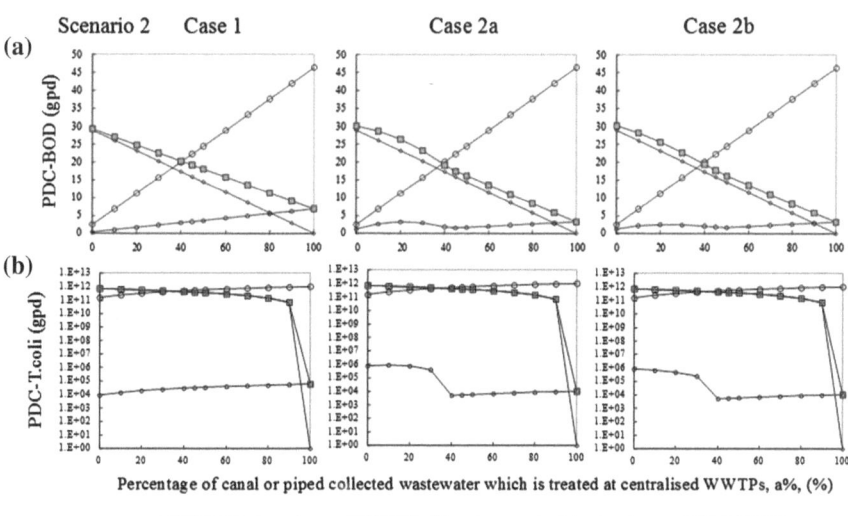

Fig. 5.8 PDC estimation results of Scenarios 1 and 2 for **a** BOD and **b** T.coli. (Tsuzuki et al. 2013b) (Copyright permission has been obtained from IWA)

existed at that time (Tsuzuki 2011). There have been at least around 10 experimental facilities in Japan at that time. Such kinds of on-site wastewater treatment system may be alternatives of on-site wastewater treatment systems in developing countries. *Huafenchi* in China would improve by some modification to resembling structure of that type of on-site wastewater treatment system.

Wastewater treatment mechanisms of *johkasou* type on-site treatment systems are mainly biological treatment. Therefore, blowers operated by electricity for aeration of aerobic chamber are necessary. Operational cost will be a challenge in developing countries. In China, willingness to pay for wastewater treatment has not been enough to afford Japanese combined *johkasou* especially for lower income people (Oh 2005). *Huafenchi* in China is included in appropriate sanitation in the MDGs sanitation indicator context. However, more improvement of people's lifestyles and ambient water quality will be necessary for the improvement. How much should we improve people's life? Advanced wastewater treatment systems are being installed to improve people's lifestyles and ambient water quality in Japan. There may be invisible barriers on the country boarders.

Relationships between pollutant discharge indicators, BOD discharge per capita (PDC-BOD) and TP discharge per capita (PDC-TP), and income indicators of developing countries in several coastal regions have been found to be inverted U-shaped curve, namely environment Kuznets curve (EKC) (Tsuzuki 2008b, 2009c). GDP per capita values at peak of the curves have been US\$ 5,050 for PDC-BOD and US\$ 11,200 for PDC-TP. The value of PDC-BOD has been consistent with the results of exiting literature, US\$ 5,000–8,000, and that of PDC-TP has been larger (Tsuzuki 2009b). TN discharge per capita (PDC-TN) decreases linearly with GDP per capita increase. The difference of peak GDP per capita values between PDC-BOD and PDC-TP can be explained by wastewater treatment technological aspect that nutrients removal is considered after achievement of organic carbon removal. One of the reasons for the linear decreasing tendency of PDC-TN has been that black water especially urine should occupy much portion of PDC-TN. Black water is usually treated with basic on-site sanitation.

Construction cost and PDC of on-site wastewater treatment systems in Asian countries are summarized based on Water Environment Partnership in Asia (WEPA) database of municipal wastewater technologies (Tsuzuki and Koottatep 2010). Construction cost and PDC are in a wide range. PDCs are estimated as 0.05–9.5 g-BOD person^{-1} day^{-1}, 0.4–5.7 g-TN person^{-1} day^{-1} and 0.02–0.82 g-TP person^{-1} day^{-1}, whereas construction cost per capita has been US\$ 4–2,942. Sanitation technologies should be suitably applied in consideration with socio-economic conditions in the area to be developed.

For municipal wastewater source control of pollutant discharge, soft measures in households have been found to be effective (Tsuzuki et al. 2012). Estimated possible pollutant discharge reduction percentages have been 38–53 % for BOD, 26–40 % for TN and 21–32 % for TP in the Yamato-gawa River Basin, Japan (Tsuzuki et al. 2009c), and 39 % for BOD, 21 % for total Kjeldahl nitrogen and 34 % for phosphate in the Bangkok Metropolitan Area, Thailand (Tsuzuki et al. 2010b), depending on lifestyles, wastewater treatment methods, and combinations of "soft interventions". Typical effective "soft interventions" are (1) to wipe plates, utensils and cookware with paper or rag before washing them in kitchen (2) to stop draining residual liquids and solids and put them in the trash can, and (3) not to drain rice washing water (Chiba Prefecture 2008; Tsuzuki et al. 2009b; Yamato-gawa River Office 2001).

Posing first priority on the MDGs water and sanitation indicators, pollutant discharge reduction should also be aimed in sanitation development to improve ambient water quality, which will benefit the society from miscellaneous aspects including water pollution control burden reduction, drinking water treatment cost

Table 5.3 Treatment efficiencies of municipal wastewater treatment processes in developed countries such as the USA and Japan

Treatment units or combinations	Removal efficiency (%)							
	BOD$_5$	COD$_{Cr}$	COD$_{Mn}$	TSS(SS)	TN	ON[a]	NH$_4$–N	TP
1 Preliminary treatment (PT)	Small[b]	Small[b]	NA[f]	Small[b]	NA[f]	Small[b]	Small[b]	Small[b]
2 Primary sedimentation	30–40	30–40	NA[f]	50–65	NA[f]	10–20	0	50–65
3 Activated sludge (conventional)	80–85	80–85	NA[f]	80–90	NA[f]	15–50	8–15	10–25
4 Tricking filter (high rate)	60–80	60–80	NA[f]	60–85	NA[f]	15–50	8–15	8–12
5 Coagulation and sedimentation after preliminary treatment or secondary treatment	40–70	40–70	NA[f]	50–80	NA[f]	50–90	0	70–90
6 Coagulation in biological treatment process	80–90	80–90	NA[f]	70–90	NA[f]	60–90	0	75–85
7 Single-stage lime addition in biological treatment	80–90	80–90	NA[f]	70–80	NA[f]	60–90	0	75–85
8 Nitrification single stage with carbonaceous BOD removal	80–95	80–95	NA[f]	70–90	NA[f]	75–85	85–95[d]	10–15
9 Denitrification separate stages suspended or attached growth[e]	small[b]	small[b]	NA[f]	small[b]	NA[f]	small[b]	small[b]	small[b]
10 Ammonia stripping	0	NA[f]	NA[f]	0	NA[f]	NA[f]	60–95	0
11 Breakpoint chlorination	NA[f]	NA[f]	NA[f]	0	NA[f]	NA[f]	80–90	0
12 Ion exchange (NH4-N)	0	0	NA[f]	0	NA[f]	0	90–95	0
13 Filtration	20–50	20–50	NA[f]	60–80	NA[f]	50–70	0	20–50
14 Carbon adsorption	50–85	50–85	NA[f]	50–80	NA[f]	30–50	NA[f]	10–30
15 Reverse osmosis	90–100	90–100	NA[f]	NA[f]	NA[f]	90–100	60–90	90–100
16 Advanced oxidation ditch	93–96	NA[f]	82–90	92–96	85	NA[f]	NA[f]	40–70
17 Coagulation in activated sludge process	92–93	NA[f]	80–85	92–95	25–30	NA[f]	NA[f]	90
18 Anaerobic-oxic activated sludge process	90–95	NA[f]	75–85	90–95	25–30	NA[f]	NA[f]	80
19 Anaerobic-anoxic-oxic activated sludge process	92	NA[f]	80–85	93	60–70	NA[f]	NA[f]	70–80
20 Coagulation and rapid filtration	50–60	NA[f]	40–60	60–70	20	NA[f]	NA[f]	60–90
21 Rapid filtration and activated carbon absorption	60–70	NA[f]	65–70	80	30	NA[f]	NA[f]	20–25

[a] Organic nitrogen

[b] BOD$_5$ or COD$_{Cr}$ removal vary if communitor and/or grid washing is used. With no communitor and/or grit washing, BOD$_5$ removal may be 0–5 % and TSS removal 5–10 %

[c] Removal is normally small and is considered to be 0 %

[d] NO$_3$-N may reach 15–20g-N m^{-3} in effluent

[e] NO$_3$-N removal rate is 85–90 %

[f] Not available

Original sources 1–15: Qasim (1985); 16–21: Japan Sewage Association (2009)

(Prepared by the author based on Qasim 1985, and Japan Sewerage Association 2009)

reduction of surface water sources, and comfortableness increases around ambient water. Japanese technologies and experiences of sanitation development together with world experiences in the fields will contribute to sanitation development in the world especially in Asia and Pacific countries.

5.7 Alternative Municipal Wastewater Treatment Systems: Technological and Institutional Options in the World and Advanced Planning Schemes in Japan

There are several options in de-centralized, on-site municipal wastewater treatment systems and low-cost sanitation (Mara 1996a, b; IDIJ 2004; Tsuzuki 2010a). There are a lot of municipal wastewater treatment methods in the world. Pollutant removal efficiencies of typical centralised WWTPs in the USA and Japan are listed in Table 5.3 as examples.

Low-cost sanitation is important concept to introduce sanitation in developing countries. Outlines of low-cost sanitation including financial, technical and administrative aspects are presented here mostly after Infrastructure Development Institute, Japan (IDIJ) (2004). IDIJ (2004) summarized the Draft Guidelines for Low-cost Sewerage Systems in Developing Countries in both Japanese and English. Some figures and tables are updated based on the recent literature and statistic data. Major sources of the IDIJ Report (2004) are as the followings.

(1) WHO/UNICEF Joint Monitoring Programme for Water Supply and Sanitation (2000) Global water supply and sanitation assessment 2000 report, 80p.
(2) Lyonnaise Des Eaux (1998) Alternative solutions for water supply and sanitation in areas with limited financial resources, prepared for UNEP.
(3) Japan Sewerage Committee for the 3rd World Water forum (2003) Water's served .... Is this all you are? pp.7–8.
(4) Ministry of Land, Infrastructure and Transport (MLIT) (2003) Outline of the estimate of annual budget requests on sewerage and wastewater in 2004, 58p., Sewerage and Wastewater Management Bureau, City and Regional Development Section.

Sanitation facilities consist of on-site facilities and off-site facilities (Fig. 5.9). Simple *johkasou* and combined *johkasou* are typical on-site wastewater treatment facilities developed in Japan. Combined *johkasou* is deployed in other countries including China, Korea, the USA, and European and Southeast Asian countries (Tsuzuki 2009b). Low-cost sewerages (Fig. 5.9) include settled sewerage, simplified sewerage and condominial sewerage. Applicable sanitation/sewerage facilities in regards to per capita consumption of water in households (10–250 L day^{-1}) are summarized in Table 5.4.

Various conditions including financial, social and technical aspects affect development of sanitation in the specific areas. For example, basic sanitation with

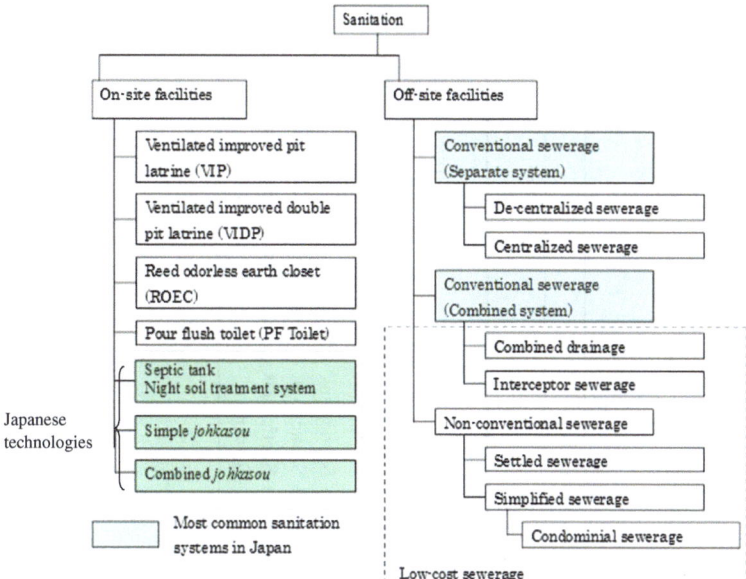

Fig. 5.9 Categorization of sanitation (Prepared by the author based on IDIJ 2004) (Copyright permissions have been obtained from IDIJ and Nova Science Publishers)

electricity, water supply and population density is summarised as shown in Fig. 5.10 (IDIJ 2004). Sanitation system can be developed along with economic development. When population density and water supply amount is small, simple pit latrine or septic tank should be appropriate sanitation facilities. When water supply amount and population density increase, appropriate sanitation facilities will be combined *johkasou*, storage pit latrine, simplified sewerage, collected toilet and drainage system, and conventional sewerage. In Japan, sanitation and wastewater treatment planning is managed in prefecture-scale (Fig. 5.11) and river basin-scale (Fig. 5.12). Zoning of municipal wastewater treatment methods is organized in prefecture scale. There are 47 prefectures in Japan. Wastewater treatment plants are transferred to other places along with time framework of the housing and industrial development in the Hyper-Flex Plan, which is considered as one of the step-wise development.

5.8 Institutional Structures and Regulations of Municipal Wastewater Treatment in Japan

Institutional structures have been established with several organisations in Japanese sanitation sector (Hashimoto 2009). These organisations correspond to several kinds of centralized and on-site wastewater treatment systems. There are

Table 5.4 Relationship between per capita consumption of domestic water and applicable sanitation/sewerage facilities

Service level for water supply (sources)	Per capita consumption (L day^{-1})			Applicable facility	
	World Bank	Duncum Mara	Lyonnaise Des Eaux	Excreta	Sullage
(1) Well etc.			50		
(2) Communal faucet	20–25	20–30	10–50	VIP latrine	Pit latrine Surface drain
(3) Yard tap	50	40–80		PF latrine + Pit latrine	Pit latrine + Sewerage
				PF latrine + Pit latrine + Sewerage	Settling box + Sewerage
				PF latrine + Settling box + Sewerage	
				PF latrine + Septic tank + Soakaway	
				PF latrine + Septic tank + Sewerage	
(4) Indoor tap	50–100		40–250	CF latrine + Septic tank + Soakaway	Settling box + Sewerage
				CF latrine + Sewerage	Sewerage

Source Infrastructure Development Institute, Japan (2004)
(*Source* IDIJ 2004) (Copyright permissions have been obtained from IDIJ and Nova Science Publishers)

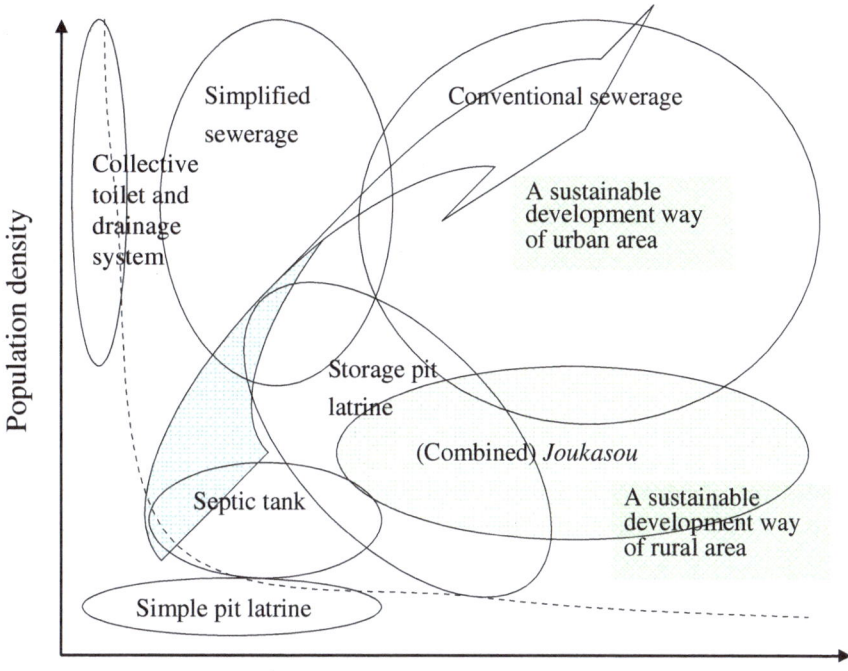

Fig. 5.10 Concept for basic sanitation with electricity, water supply and population density (Modified from IDIJ 2004) (Copyright permissions have been obtained from IDIJ and Nova Science Publishers)

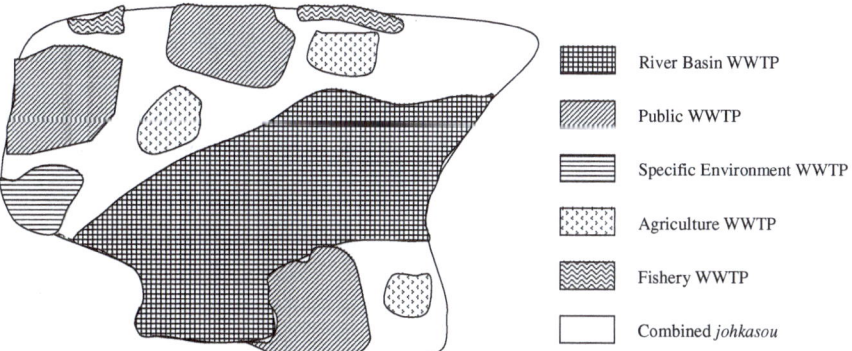

Fig. 5.11 A conceptual figure of wastewater treatment prefecture plan of Japan (Copyright permission has been obtained from IDIJ)

Fig. 5.12 A conceptual
figure of the River Basin-
Wide Planning of Sewerage
Systems in Japan. (Prepared
by the author based on IDIJ
2004)

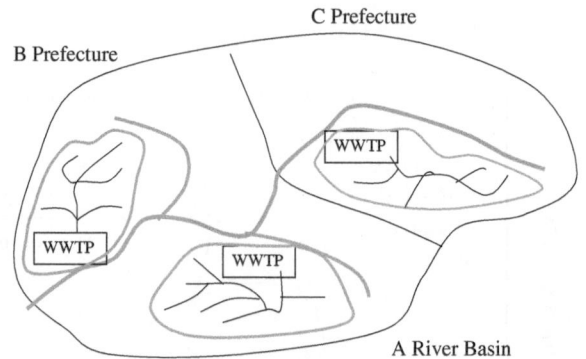

multiple organisations for certain wastewater treatment systems, e.g. for public
WWTPs, an organisation deals with technology research and development,
another organisation deals with information dissemination, another organisation
deals with piped-collection systems and so on. Japan Sewerage Works Agency is
funded by local governments and maintains advancement of sewerage works
engineering (Japan Sewerage Works Agency 2010). Japan Sewerage Works
Association works for information dissemination on sewerage works (Japan
Sewerage Works Association 2009). Japan Environmental Sanitation Centre pur-
sues environmental protection and conservation especially in the fields of solid
wastes, night soil treatment system and acid rain (Japan Environmental Sanitation
Centre 2009). Japan Education Centre for Environment Sanitation focuses on
johkasou technology including simple and combined *johkasou* and conducts
education and research for training technicians of installation, maintenance and de-
sludging of the on-site treatment systems (Ogawa 2006; Japan Education Centre
for Environment Sanitation 2010). Such kind of complicated structure of organisations in the fields of municipal wastewater treatment has been developed in
Japanese sanitation sector for several decades.

For infectious disease prevention, disinfection of effluent is necessary especially for biological wastewater treatment effluent. Chlorination using chlorine
tablets are sometimes applied for biologically treated wastewater of *johkasou*.
Investigation on the relationship between solubility and disinfection efficiency of
chlorine tablets have been conducted (Yahya 1994). Other disinfection methods
such as ultraviolet and ozone disinfection are also applied for effluent of centralized and on-site wastewater treatment systems. For basin wide WWTPs, local
government WWTPs and specific environment WWTPs, there have been 1964
WWTPs with chlorination using chlorine tablets and liquid type chlorine, 110
WWTPs with ultraviolet disinfection, and 28 WWTPs with ozonation at the end of
March 2006 in Japan (Japan Sewerage Works Association 2007).

For pollutant removal efficiencies of combined *johkasou*, regulations of small
scale *johkasou* including combined *johkasou* and simple *johkasou* has been firstly
established as double standards: (1) BOD removal rate of more than 65 % and

effluent BOD is less than 90 g-BOD m^{-3} in the Notification of Ministry Construction (Currently, Ministry of Land, Infrastructure, Transportation and Tourism),[7] and (2) BOD removal rate of more than 90 % and effluent BOD is less than 20 g-BOD m^{-3} in the Notification of Architecture Guidance Section of Housing Bureau in the Ministry[8] in 1988 (Kitao 2006). The former Notification has subjected both combined and simple *johkasou*.

Pollutant removal efficiencies of small scale *johkasou* have been increased by research and development and currently as large as those of centralized WWTPs. Building Standards Act has been revised to include performance evaluation system into the existed structure based evaluation system in 1998.[9] Percentage of newly installation of performance based combined *johkasou* has been increased from around 20 % in 1998 to 93 % in 2005. National Combined *Johkasou* Promotion Local Governments Union classified four categories among performance based evaluation standards of combined *johkasou* by effluent concentrations, 1) BOD removal type: 20 g-BOD m^{-3}, 2) nitrogen removal types (three types): 20 g-BOD m^{-3} and 20 g-total nitrogen (TN) m^{-3}; 10 g-BOD m^{-3} and 20 g-TN m^{-3}; and 10 g-BOD m^{-3} and 10 g-TN m^{-3}, 3) nitrogen and phosphorus removal type: 10 g-BOD m^{-3}, 10 g-TN m^{-3} and 1 g-total phosphorus (TP) m^{-3}, 4) BOD advanced treatment type: 5 g-BOD m^{-3} and 10 g-TN m^{-3}. Pollutant removal ratios of combined *johkasou* with these performance based standards especially for nutrients (nitrogen and phosphorus) removal types are comparable to those of centralized WWTPs with tertiary or advanced treatment.

Operation of *johkasou* is conducted by registered vendors and operators. The numbers of the vendors, operators and technicians have been 38 *johkasou* manufacturers, 65 specified examination agencies, 77,728 *johkasou* installation workers, 9,087 vendors for *johkasou* operation and maintenance, 63,104 *johkasou* operators, 4,806 vendors for *johkasou* de-sludging, 4,205 *johkasou* de-sludging technicians and 2,533 *johkasou* inspectors at the end of March 2005 (Ogawa 2006). These kinds of structures are necessary to construct and maintain *johkasou* on-site municipal wastewater treatment systems in Japan.

References

Basin Wide Sewerage Works Comprehensive Planning System Design Council, Japan (2008) Survey on basin wide sewerage works comprehensive planning: guidelines and explanation. Japan Sewerage Works Association, p 285 (in Japanese)

Chiba Prefecture (2008) Action plan to improve environment of Tokyo Bay by everyone. http://pref.cihba.lg.jp/syozoku/e_suiho/1_sido/suishinkeikaku/plan/suiplanup.html (in Japanese)

---

[7] Notification 1292 of Ministry Construction on 8 March 1988.

[8] Notification of Architecture Guidance Section of Housing Bureau in the Ministry on 8 March 1988.

[9] Act No. 100 in 1998 established on 12 June 1998.

Franceys R, Pickford J, Reed R (1992) A guide to the development of on-site sanitation. WHO, p 229. http://www.who.int/water_sanitation_health/hygiene/envsan/onsitesan/en/

Fujimoto C (1988) Pollutant load of rivers flowing into inbanuma and teganuma marshes, basic units of domestic wastewater. Chiba water quality conservation institute annual report of Syowa 62(1987):89–98 (in Japanese)

Fujimura Y (1996) Pollutant load basic units of domestic wastewater and discharge rate of jokaso. Chiba prefecture water quality conservation institute annual report of Heisei 7(1995):33–38 (in Japanese)

Fujimura Y, Nakajima A (1998) Domestic wastewater pollutant loads and combined jokaso. Inbanuma marsh nature and culture 5:27–34 (in Japanese)

Gaulke LS (2006) On-site wastewater treatment and reuses in Japan. Proc ICE Water Manage 159:103–109

Hashimoto K (2009) Sanitation issues in the developing countries. J Jpn Soc Water Environ 32(9):2–6 (in Japanese)

Infrastructure Development Institute, Japan (2004) Guidelines for low-cost sewerage systems in developing countries (draft), p 270 (in Japanese and in English)

Ishii K, Yamada K (1990) Revolution of municipal wastewater treatment: escape from river environment deterioration. Fujiwara, Tokyo,p 222 (in Japanese)

Japan Education Centre for Environment Sanitation (2010) Japan education centre for environment sanitation homepage http://www.jeces.or.jp/en/about/index.html

Japan Environmental Sanitation Centre (2009) Japan Environmental Sanitation Centre homepage http://www.jesc.or.jp/en/index.html

Japan Sewerage Association (2009) Guidelines of planning and designing and their explanations sewerage facilities 2009, Tokyo, Japan (in Japanese)

Japan Sewerage Committee for the 3rd World Water forum (2003) Water's served …. Is this all you are?, pp 7–8

Japan Sewerage Works Association (2007) Sewerage works statistics in Japanese fiscal Year 2005: public administrative volume. Japan Sewerage Works Association (in Japanese)

Japan Sewerage Works Association (2009) Japan sewerage works association homepage http://www.jswa.jp (in Japanese)

Japan Sewerage Works Agency (2010) Japan Sewerage works agency homepage. http://www.jswa.go.jp/english/index.html

Jiawkok S, Koottatep T (2006) Assessment of on-site sanitation systems in peri-urban communities by using selected sustainability indicators. Msc thesis, Asian Institute of Technology, Pathumthani, Thailand

Kitao T (2006) Structure and maintenance of small johkasou. Japan Education Centre for Environment Sanitation, Tokyo, p 196 (in Japanese)

Lyonnaise Des Eaux (1998) alternative solutions for water supply and sanitation in areas with limited financial resources, prepared for UNEP

Mara D (1996a) Low-cost sewerage. Wiley, London, p 238

Mara D (1996b) Low-cost urban sanitation. Wiley, London, p 240

Ministry of Agriculture, Forestry and Fisheries, Japan (2008b) Leaflet: agriculture village WWTPs http://www.maff.go.jp/j/press/nousin/sousei/pdf/080528-03.pdf (in Japanese)

Ministry of Land, Infrastructure, Transportation and Tourism, Japan (2009) Sewerage works data room http://www.mlit.go.jp/crd/crd_sewerage_tk_000104.html (in Japanese)

Ministry of Agriculture, Forestry and Fisheries, Japan (2009a) Press release: municipal wastewater treatment percentage in fishery village at the end of Japanese Fiscal Year 2008 http://www.jfa.maff.go.jp/j/press/bousai/090930.html (in Japanese)

Nakanishi J, Okinoto T (1982) Theory of municipal wastewater treatment planning: environment impact assessment of municipal wastewater treatment planning at Komagane City, Japan. Musashino-Shobo, Tokyo, p 222 (in Japanese)

Ogawa H (2006) Domestic wastewater treatment by johkasou systems in Japan. Side Event: potential for Johkasoh (Private Wastewater Treatment Tank) in Asia, Asia 3R Conference 30

Oct 1, Nov 2006, Tokyo http://www.env.go.jp/recycle/3r/en/asia.html, accessed on 10 Apr 2010

Oh H (2005) Research on johkasou technology transfer in China: case study of Harbin City. Master Thesis of Toyo University, Japan (in Japanese)

Qasim SR (1985) Wastewater treatment plants: planning, design and operation, CBS College Publishing, Holt, p 726

Serageldin I (1994) Water supply, sanitation, and environmental sustainability: the financing challenge. The World Bank, Washington, p 42 http://www-wds.worldbank.org/external/default/WDSContentServer/WDSP/IB/1994/11/01/000009265_3970716143355/Rendered/PDF/multi_page.pdf

Sewerage and Wastewater Management Department, City and Regional Development, Ministry of Land, Infrastructure and Transport (MLIT) (2003) Outline of the estimate of annual budget requests on Sewerage and Wastewater in 2004, p 58

Sinsupan T, Koottatep T (2004) Material flux analysis (MFA) for planning of domestic wastes and wastewater management: case study in Pak Kret Municipality, Nonthaburi, Thailand. AIT Master Thesis, Bangkok

Suharyanto MJ (2009) Comparative study on integrated wastewater management system model for developing countries under rapid urbanization. In: 7th international symposium on Southeast Asian Water Environment, Bangkok, 28–30 Oct 2009

Tilley E, Lüthi C, Morel A, Zurbrügg C, Schertenleib R (2008) Compendium of sanitation systems and technologies, Swiss Federal Institute of Aquatic Science and Technology (Eawag), Dübendorf 158

Tsuzuki Y (2006) An index directly indicates land-based pollutant load contributions of domestic wastewater to the water pollution and its application. Sci Total Environ 370(2–3):425–440

Tsuzuki Y (2008a) Erratum to Tsuzuki Y (2006) An index directly indicates land-based pollutant load contributions of domestic wastewater to the water pollution and its application, Science of the Total Environment, 370:425–440, Sci Total Environ 395:50–50

Tsuzuki Y (2008b) Relationships between water pollutant discharges per capita (PDCs) and indicators of economic level, water supply and sanitation in developing countries. Ecol Econ 68:273–287

Tsuzuki Y (2009a) Comparison of pollutant discharges per capita (PDC) and its relationships with economic development: an indicator for ambient water quality improvement as well as the Millennium Development Goals (MDGs) sanitation indicator. Ecol Ind 9:971–981

Tsuzuki Y (2009b) Wastewater discharge regulations and new development of on-site wastewater treatment systems in Thailand. J Johkasou—Small Domest Wastewater Treat Syst—(Gekkan Johkasou) 393:37–40 (in Japanese)

Tsuzuki Y (2010a) Chapter 5: Domestic wastewater pollutant discharge and pollutant load water quality in the ambient water in developed and developing countries, 125–164, in Kudret Ertuð and Ilker Mirza eds. Water Quality: Physical, chemical and biological characteristics, Nova Science Publishers, Inc., p 277, ISBN: 978-1-60741-633-3

Tsuzuki Y (2011) Chapter 6: Sanitation development and roles of Japan, 179–202. In: McMann JM (ed) Potable water and sanitation, Nova Science Publishers, Inc., New York, 266p. ISBN: 978-1-61122-319-4

Tsuzuki Y (2012) Linking sanitation and wastewater treatment: from evaluation on the basis of effluent pollutant concentrations to evaluation on the basis of pollutant removal efficiencies. Water Sci Technol 65(2):368–379

Tsuzuki Y, Koottatep T (2010) Municipal wastewater pollutant discharge indicator estimation and water quality prediction in pak kret district. Bangkok, Thailand, Journal of Water and Environment Technology, Japan Society on Water Environment 8(1):51–75

Tsuzuki Y, Koottatep T, Rahman MM, Ahmed F (2008a) Water quality in the ambient water and domestic wastewater pollutant discharges in the developing countries: possibility of combined johkasou exports. Joukasou Kenkyu (J Domest Wastewater Treat Res) 20(1):1–13 (in Japanese with English abstract)

Tsuzuki Y, Koottatep T, Ahmed F, Rahman MM (2008b) Domestic wastewater pollutant discharge and pollutant load in the tidal area of the ambient water in developing countries: survey results in autumn and winter in 2006. J Global Environ Eng 13:121–133 (Japan Society of Civil Engineers) http://www.wsscc.org/resources/resource-publications/water-quality-and-pollutant-load-ambient-water-and-domestic

Tsuzuki Y, Koottatep T, Rahman MM (2009a) Water quality profiles of the tidal rivers and canal in per-urban of Bangkok, Thailand, and Dhaka, Bangladesh, focusing on the water quality transition in coastal areas. J Jpn Soc Water Environ 32(1):47–52 (in Japanese with English abstract)

Tsuzuki Y, Koottatep T, Wattanachira S, Sarathai Y, Wongburana C (2009b) On-site treatment systems in the wastewater treatment plants (WWTPs) service areas in Thailand: scenario based pollutant loads estimation. J Global Environ Eng Jpn Soc Civ Eng 14:57–65

Tsuzuki Y, Fujii M, Mochihara Y, Matsuda K, Yoneda M (2009c) 12th International river rymposium, Brisbane, Australia, Sep 2009. http://www.riversymposium.com/index.php?page=2009

Tsuzuki Y, Koottatep T, Jiawkok S, Saenpeng S (2010) Municipal wastewater characteristics in Thailand and effects of soft intervention measures in households on pollutant discharge reduction. Water Sci Technol 62(2):231–244

Tsuzuki Y, Yoneda M, Tokunaga R, Morisawa S (2012) Quantitative evaluation of effects of the soft interventions or cleaner production in households and the hard interventions: a social experiment programme in a large river basin in Japan. Ecol Ind 20:282–294

Tsuzuki Y, Koottatep T, Sinsupan T, Jiawkok S, Wongburana C, Wattanachira S, Sarathai Y (2013a) A concept in planning and management of on-site and centralised municipal wastewater treatment systems, a case study in Bangkok, Thailand I: pollutant discharge indicators and pollutant removal efficiency functions. Water Sci Technol 67(9):1923–1933

Tsuzuki Y, Koottatep T, Sinsupan T, Jiawkok S, Wongburana C, Wattanachira S, Sarathai Y (2013b) A concept in planning and management of on-site and centralised municipal wastewater treatment systems, a case study in Bangkok, Thailand II: Scenario-based pollutant load analysis. Water Sci Technol 67(9):1934–1944

UNDP (2007) Statistics of the Human Development Report http://hdr.undp.org/en/statistics/

WHO/UNICEF Joint Monitoring Programme for Water Supply and Sanitation (2000) Global water supply and sanitation assessment 2000 Report, p 80

Yahya AS (1994) Characteristics of chlorine tablet solubility and their disinfection efficiency of *johkasou* effluents. Bull Inst Public Health Jpn 43(3):316–319 (in English with Japanese abstract)

Yamato-gawa River Office, Ministry of Land, Infrastructure, Transportation and Tourism (2010) Social experiment program on municipal wastewater pollutant discharge reduction http://www.kkr.mlit.go.jp/yamato/drainage/index.html (in Japanese)

Chapter 6
Water and Sanitation in Developing Countries

There are many kinds of technological options in water supply, sanitation and wastewater treatment. The water MDGs may be achievable in 2015, however, the sanitation MDGs may be hard to be achieved. Therefore, low-cost sanitation is necessary especially in the critical regions. In this chapter, some alternatives for the conventional wastewater treatment and sanitation systems in the concept of low-cost sanitation are explained to consider the methods to develop appropriate municipal wastewater treatment systems. Some typical wastewater treatment planning schemes in Japan will also give some suggestions to developing countries especially with larger MDG sanitation indicator.

6.1 Water and Sanitation in Developing Countries

Sanitation and water quality chemical analyses have been developed together since around 1900 in the world (Table 6.1). The basic wastewater treatment systems including filtration, septic tank and activated sludge have been developed in 1800–1910s. The first Standard Methods has been published in the 1900s for general water quality parameters. Technologies and concepts of low-cost sanitation have developed since around the 1950s (IDIJ 2004). Some of the notable experiences of the low cost sanitation have been from Zambia, Brazil and Pakistan in 1960–1983. Several international and national organizations including World Bank, United Nations Centre for Human Settlements (UNCHS or Habitat), Water Pollution Control Federation (WPCF), US-EPA have made significant contributions in low-cost sanitation development. The knowledge and information on low-cost sanitation and sewerage have been summarised in the middle 1990s (Mara 1996a, b).

Cost of sanitation and sewerage should be considered with the places and time because of technological development, and financial and social aspects. Table 6.2 shows examples of the water and sanitation/sewerage costs. When we consider the costs, we should also consider possible charge for and benefit of sanitation facilities. One of the most commonly applied money collection methods of

Y. Tsuzuki, *Pollutant Discharge and Water Quality in Urbanisation*,
SpringerBriefs in Water Science and Technology,
DOI: 10.1007/978-3-319-04756-0_6, © The Author(s) 2014

Table 6.1 Brief history of sanitation, chemical analytical method and low-cost sanitation in the world (Prepared by the author based on Mara 1996a, b; IDIJ 2004; Shifrin 2005; and Lofrano and Brown 2010)

Year	Sanitation	Evolution of environmental analytical chemistry	Low-cost sanitation	
			Organization, author or place	Title or conference
BC800–AD476	Roman times			
476–1800	Sanitary dark age			
1800–1914	Developing the basic treatment process			
1860	Filtration process			
1985	Septic tank			
1902	Imhoff tank			
1905	Radial flow tank			
1900s		1st standard methods: General parameters		
1910s	Activated sludge	3rd standard methods: General parameters BOD and DO methods improvements Phenol > 10 mg/l		
1914–1965	The age of process development			
1930s		8th standard methods: General parameters BOD, COD and DO methods improvements Phenol > 10 mg/l Petroleum hydrocarbon fractions		
1940s		Infrared (IR) method for some inorganics > 500 mg/l Benzene >100 mg/l DO by mercury electrode UV photometry for some organics > 30 mg/l Total petroleum		
1950s	Constructed wetland (CW) Nitrogen removal	Potable DO Introduction of mass spectrometry (MS)		

(continued)

Table 6.1 (continued)

Year	Sanitation	Evolution of environmental analytical chemistry	Low-cost sanitation	
			Organization, author or place	Title or conference
Late 1950s		Introduction of atomic adsorption spectroscopy (AAS) IR methods advances Column chromat	L. J. Vincent (Zambia)	Development of settled sewerage system
1960			Kafue, Zambia	First installation of settled sewerage system
1960s	Rotating biological contact (RBC) reactor	11th Standard Methods: Wet chemical inorganics GC advances Fluorescence spectroscopy IR advances PCB and DDT detection by GC First commercial AAS		
1965–2000	Process refinement			
1970s	Upflow anaerobic sludge blanket (UASB) Phosphorus removal	PCB Aroclor and separation refinement GC refinements (cleanup, flame ionization, electron capture) MS and Nuclear magnetic resonance refinements Inductive Coupled Plasma Aromic Emission Spectraphotometry ICP First EPA manual on organic and trace metal analysis		
Early 1980s			Nagar city, Rio Grande de Notre, Brazil	Development of condominal sewerage system

(continued)

Table 6.1 (continued)

Year	Sanitation	Evolution of environmental analytical chemistry	Low-cost sanitation	
			Organization, author or place	Title or conference
1980s	Membrane biological reactor (MBR) Sequencing batch reactor (SBR)	First EPA manual on hazardous waste analysis		
1983			Karachi, Pakistan	Commencement of Orangi pilot project (ORP)
1985			World Bank TAG	Technical note on small-bore sewerage
1986			United Nations Centre for Human Settlements (Habiat) (UNCHS)	The design of shallow sewer system
1986			Water Pollution Control Federation (WPCF), US	Alternative sewer systems (MOP No. FD-12)
1990s	Moving bed biofilm reactor (MBBR)	20th standard methods: 350 separate methods Dioxin analysis > 1 part per quadrillion		
1991			Environmental Protection Agency (EPA), US	Manual—alternative wastewater Collection system
1994			UNDP-World Bank Water and sanitation program	Simplified sewerage: Design guidelines
19–21 July 1995			University of Leeds, UK	International conference on low-cost sewerage
1996			Duncan Mara (UK)	Low-cost sanitation
			Duncan Mara (UK)	Low-cost sewerage

Table 6.2 Initial investment cost per capita of water and sanitation (Prepared by the author from IDIJ 2004; Hutton and Haller 2004. Copyright permissions have been obtained from IDIJ and Nova Science Publishers)

Improvement		Initial investment cost per capita			Initial investment cost per capita or household (HH)			
		Africa, US$ Year	Asia, 2000/ capita	LA&C	World Bank, US$/ HH	Thailand, Baht/ unit	Indonesia, Rp/ capita	Brazil, US$/ capita
Water	House connection	102	92	144				
	Standpost	31	64	41				
	Borehole	23	17	55				
	Dug well	21	22	48				
	Rainwater	49	34	36				
	Disinfection at point of use	0.13	0.094	0.273				
Sanitation	Sewer connection	120	154	160				
	Small bore sewer	52	60	112				
	Septic tank	115	104	160		4,000–9,000		
	Pour-flush latrine	91	50	60	75–225	2000		
	VIP latrine	57	50	52	50–150			
	Simple pit latrine	39	26	60				
	Johkasou					21,500		
Sewerage	Conventional						$80–120 \times 10^3$	240–390
	Settled						$70–90 \times 10^3$	
	Simplified							170–240
	Condominal							65–105

Table 6.3 Alternative financial solutions in water supply (Prepared by the author based on IDIJ 2004; Lyonnaise Des Eaux 1998. Copyright permissions have been obtained from IDIJ and Nova Science Publishers)

Payment methods	Description	Connection meter	Infrastructure (secondary network)	Water tariff	Work on private property
Participation (labor)	The user does the excavation (or all the work) him/herself, in return for a price reduction	x[a]	x[a]	x[a]	
Grant	Municipality supply materials for connection. The user may or may not provide labor	x[a]	x[a]	x[a]	x[a]
Welfare connection	For a particular population category, the connection cost or cost difference is paid by municipality or management body. There is difficulty in finding the recipients	x[a]	x[a]		
Work funds	In Casablanca, the work fund was provided for the connection charges of the users and specifically defined works				
Micro-loan	Micro-loans were originally developed to encourage for the micro-enterprise establishments. Micro-loans are provided to give disadvantaged population access to credit	x[a]	x[a]		x[a]
Tontine, pooled investment, and security	Mutual aids developed by some communities. Such private sectors are usually not included in the development schemes, however, knowledge of these schemes in the communities will help smooth implication of development projects		x[a]	x[a]	x[a]

[a] Example of application

sanitation facilities is pricing water supply. In this sense, financial aspects of water supply are important even from cost recovery perspective of sanitation especially in developing countries (Table 6.3).

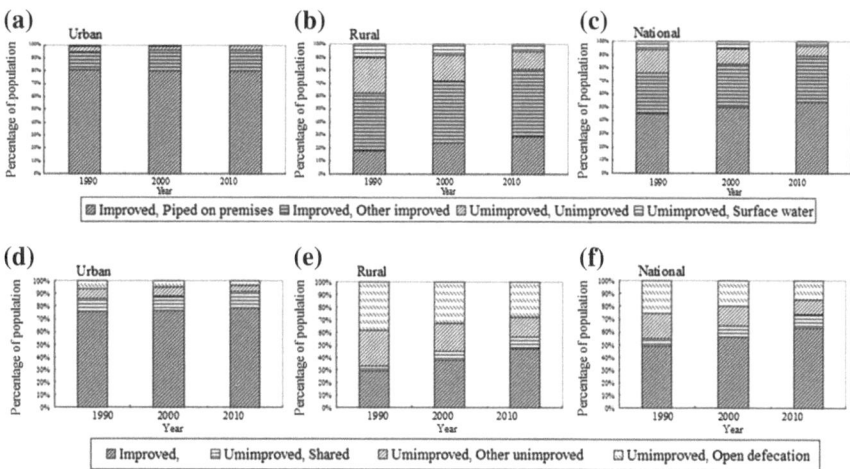

Fig. 6.1 Percentage of population with improved and unimproved water supply in **a** urban **b** rural and **c** national areas, and improved and unimproved sanitation in **d** urban **e** rural and **f** national areas (Prepared by the author based on UNICEF and WHO 2012)

Worldwide situations of water and sanitation have been disseminated through the World Bank or the United Nations organizations in these days especially in related to the MDG goals including WHO and UNICEF. Percentage of population with unimproved water supply in national areas (total areas) has been 24 % in 1990, 17 % in 2000 and 11 % in 2010 (Fig. 6.1a–c). Percentage of urban population has been increased from 43 % in 1990, to 46 % in 2000 and 51 % in 2010. It means that the MDGs water supply indicator has been achieved in 2010 in the whole world while there are still specific regions and some rural areas where the indicator may be hard to be achieved by 2015.

On the contrary, percentage of population with unimproved sanitation in national areas has been 51 % in 1990, 44 % in 2000 and 37 % in 2010 (Fig. 6.1d–f). The percentage should be decreased smaller than 25.5 % by 2015 in the MDGs sanitation indicator, and it is considered to be severe target.

Among regions, large percentages of unimproved water supply in 2010 are found to be 46 % in Oceania, and 39 % ub Sub-Saharan Africa and those of unimproved sanitation are found to be 70 % in Sub-Saharan Africa, 59 % in Southern Asia, 45 % in Oceania, 34 % in East Asia, and 31 % in Southeast Asia (Fig. 6.2).

Total water supply investment in 1990–2000 in the world has been US$12.5 billion in 1990–2000, and that of sanitation has been US$3.1 billion (Fig. 6.3).

Purposes of most of the appropriate sanitation facilities in the MDG perspective are to prevent infectious diseases and to improve lifestyles. Pollutant discharge reductions with some basic sanitation facilities are not as large as conventional

(a)

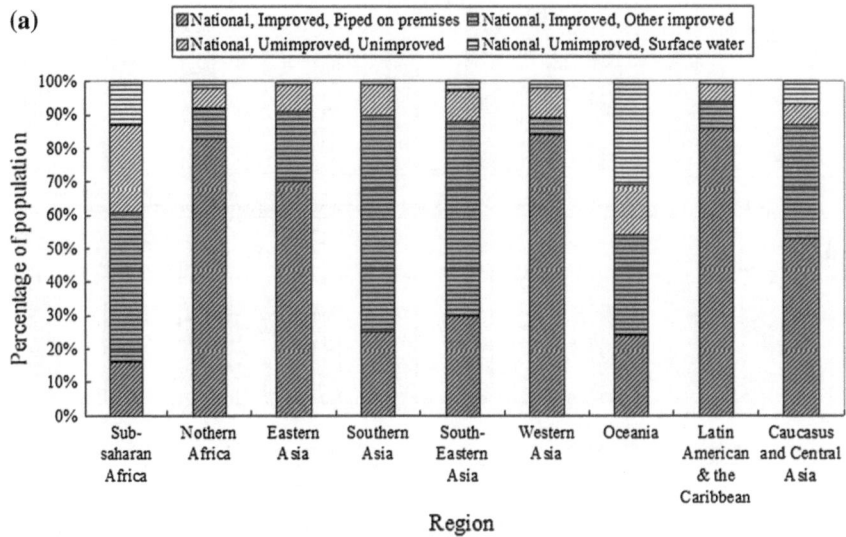

(b)

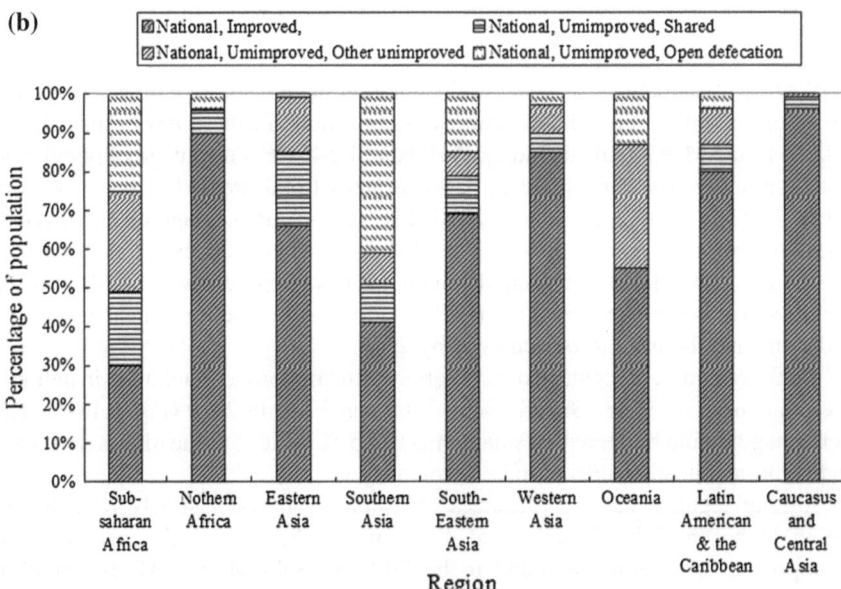

Fig. 6.2 Service coverage percentage with water supply and sanitation by regions in 2010 (Prepared by the author based on UNICEF and WHO 2012)

sewerage or up to date sanitation systems in developed countries including Japan. More improvement should be considered for sanitation in the countries with large MDG sanitation indicators such as Thailand, 99–100 %, to improve the ambient water quality with pollutant discharge reduction (Fig. 6.4) (Tsuzuki et al. 2008a, b).

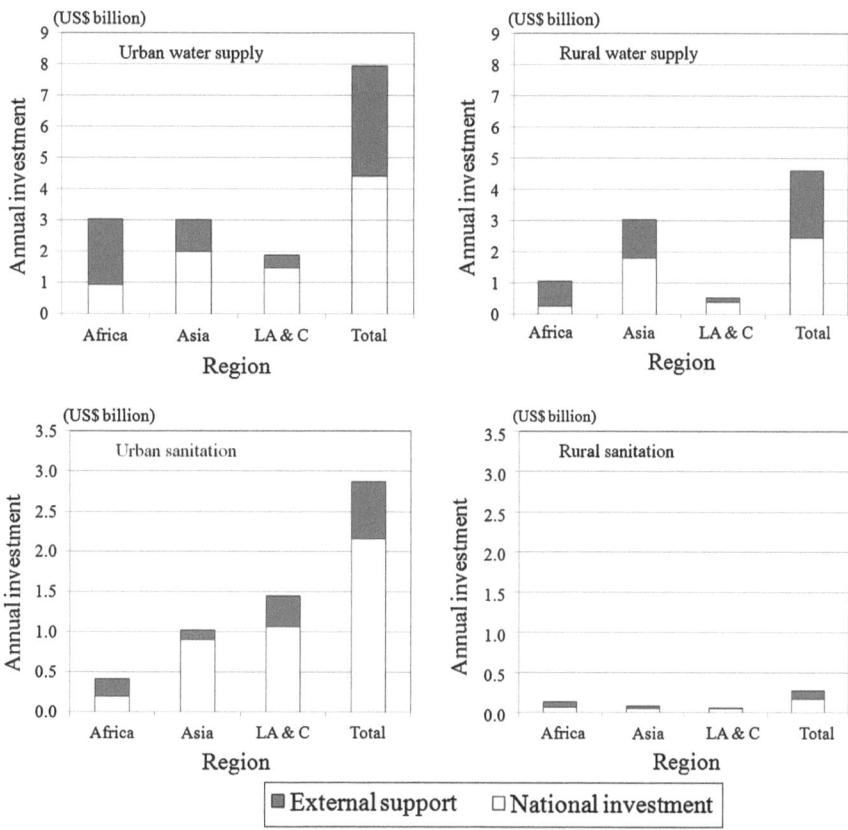

Fig. 6.3 Investments in water supply and sanitation (1990–2000) (Prepared by the author based on WHO/UNICEF Joint Monitoring Programme for Water Supply and Sanitation 2000)

6.2 Conceptual Development in Sanitation Sector

For sanitation, there are many purposes including hygiene, infectious diseases and mortality (Bartram et al. 2005; Fewtrell et al. 2005), children growth (Checkley et al. 2004) and food processing hygiene (Norman and Gravani 2006). Cost of food sanitation has been estimated as US$229–610 million for E. coli O157:H7 outbreak for beef parties industry of the fast food restaurant, US$263 million for benzene contamination in a popular brand imported bottled water, and US$10–50 million for inadvertent defective glass of export bottles by a European beer maker.

Every year, 1.8 million people die from diarrhoeal[1] diseases including cholera, among which 90 % are children under 5, mostly in developing countries (World

---

[1] Both diarrhoea and diarrhoea are used in this manuscript mostly based on the source references.

Fig. 6.4 Resources
necessary for sanitation
development (Tsuzuki 2011)
(Copyright permission has
been obtained from Nova
Science Publishers)

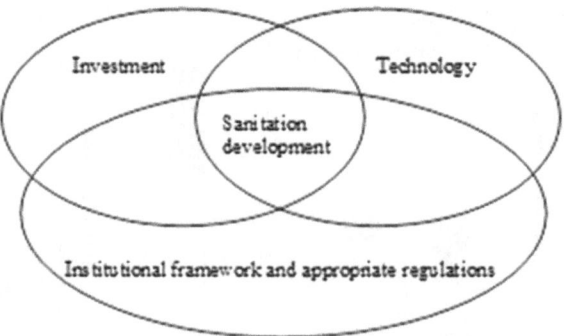

Health Organisation 2004). Unsafe drinking water and inadequate sanitation and hygiene have been attributed to 88 % of diarrhoeal disease. Approximately 3.1 % of 1.7 million deaths and 3.7 % of disability-adjusted life years (DALYs) of 54.2 million in the world have been estimated to be attributable to unsafe drinking water and inappropriate sanitation and hygiene (Water Supply and Sanitation Collaboration Council and World Health Organisation 2005). The daily mortality number, about 5,000 mortality per day, corresponds to passengers of 10 large-size airplanes (Harada 2010).

After the Water and Sanitation Decade in 1981–1990, increasing percentages of people with safe drinking water supply and appropriate sanitation have been targeted in the water and sanitation sectors in the world. The percentage without access to improved water sources has been 23 % in 1990, decreased linearly to 13 % in 2008, and estimated to reach to 9 % in 2015, the targeted year of the Millennium Development Goals (MDGs) (WHO/UNICEF Joint Monitoring Programme on Water Supply and Sanitation 2010). The water supply MDGs target of 12 % is estimated to be achieved with continuing efforts in water sector. On the contrary, the MDGs sanitation indicator has been still 39 % in 2008 and projected sanitation indicator is 36 % in 2015 whereas the MDGs sanitation target aims for 23 %. The MDGs sanitation sector is considered to be hard to be achieved. There should be more efforts in sanitation sector to prevent infectious diseases and to improve people's lifestyles and ambient water quality. After the Water and Sanitation Decade, conceptual development has been achieved in the sanitation sectors (Table 6.4).

Technical options for on-site sanitation and financial and institutional requirements for sanitation development especially for developing countries have been comprehensively summarised (Franceys et al. 1992). The technical options include open defecation, shallow pit, simple pit latrine, borehole latrine, ventilated pit latrine, pour-flush latrine, single or double pit, composting latrine, septic tank, and aqua-privy. Removal systems for excrete included in their summary have been overhung latrine, bucket latrine, vaults and cesspits and sewerage. Planning and designing of on-site sanitation have also been discussed including designs of pit latrine, septic tank, aqua-privy and composting toilets, and planning schemes from project definition to institutional, economic and financial factors.

Table 6.4 Conceptual development in sanitation sector after 1990

Parameters Resources for sanitation	Benefit of sanitation	Explanation	Reference
Labour, building materials, water, land use, timing of costs, design life of the system, emptying and disposal, groundwater pollution, sullage disposal, wastewater reuse, and administrative management	Enhanced privacy, convenience for the users, environmental protection, reduction and anticipated eventual elimination of excrete-related disease, increase in productive life expectancy, increase in work capacity, reduction of demand for medical facilities and drugs, reuse of composted or digested excrete for agricultural purposes, and production of biogas for energy needs	Standard duration of cost analyses have been recommended as 10–15 years. TACH[a] should be estimated for the standard duration time including all the necessary financial factors. The least-cost analysis approach has been appreciated	Franceys et al. (1992) WHO
Financing who pays for water and sanitation projects; pro-rich or pro-poor; community-based structures for water and sanitation sector investments; tariff paid by the customers and connection rate; and receipts and expenditures for industry and public		Water sector projects have been pro-poor, i.e. poor people have been beneficial, in countries with large per capita Gross National Product (GNP) and pro-rich, i.e. rich people have been beneficial, in low-income countries. On the contrary, sanitation sector projects have been pro-rich in all the analysed countries	Serageldin (1994) World Bank
	RR[b] and DALY[c] of diarrhea and other diseases	DALY[c] has been estimated based on the regional RR[a] data using exposure scenarios. Disease burden from water, sanitation and hygiene to be 4.0 % of all deaths and 5.7 % of the total diseases burden (in DALYs) in the world in consideration with diarrheal disease, schistosomiasis, trachoma, ascariasis, trichuriasis and hookworm diseases	Prüss et al. (2002) Environ Health Perspect

(continued)

Table 6.4 (continued)

Parameters Resources for sanitation	Benefit of sanitation	Explanation	Reference
	Access to water and sanitation infrastructure, availability of water, improved health behaviours, diarrhea rates in children under five, and under-age-five mortality rate	Distance to water sources and under-age-five mortality have been reduced significantly in the two areas, and sanitation facilities improvement has been found in one of the two treatment area. Drinking water quality did not improve significantly	Newman et al. (2002) World Bank Economic Review
	Access to water and sanitation infrastructure, improved health behaviours, diarrheal rates in children under five, anthropomorphic measures in children under five, and expenditure levels on water, and income	Water supply and sanitation intervention benefits on child health and income will be calculated using a difference-in-difference estimator to measure differences between the villages with the Jalswarajya Project (treatments) and non-project villages (controls)	Poulos et al. (2006) (Original source: Pattanayak et al. 2005) World Bank
Coherence, predictability and credibility, and legitimacy and accountability		Good design will incorporate lessons from international experience including working with the existing organisational framework, creating an appropriate role for politics, limiting the discretion given to regulatory decision makers, and trading off sophistication in favor of simplicity. Regulations should be prepared to be appropriate for unique social and economic conditions by country	Ehrhardt et al. (2007) World Bank

[a] Total annual cost per household
[b] Relative risk
[c] Disability-adjusted life year

A comparison on the financial sources of the World Bank-financed water and sanitation projects has been conducted (Serageldin 1994). Whereas there have been regional differences of the financial sources among the regions, overall percentage of projects financed by the internal cash generation has decreased by time: 34 % in 1988, 22 % in 1989, 18 % in 1990 and 10 % in 1991. In the "new agenda", it has been important to consider not only total budget but also who benefits from subsidized from water and sanitation services. Analysis results of water sector projects in the Latin American countries revealed that water sector projects are pro-poor, i.e. poor people are beneficial, in countries with large per capita Gross National Product (GNP) and pro-rich, i.e. rich people are beneficial, in low-income countries. On the contrary, sanitation sector projects are pro-rich in all the analysed countries. Community based organisational structures for water and sanitation sector development have also been discussed. These are Ruhverband in the Ruhr area in Germany, nation-wide River Basin Financing Agencies in France, and condominal sewerage system in Brazil (Serageldin 1994).

The two textbooks on low-cost wastewater treatment and sanitation have been prepared to improve understanding of sanitation development which requires a mixture of resources including technological, financial and institutional resources after experiences of the Water and Sanitation Decade in 1981–1990 (Mara 1996a, b). Broad socio-economic aspects of urban sanitation development have been discussed based on the study results of the Stockholm Environment Institute conducted on environment burdens in Accra, Ghana, Jakarta, Indonesia, and Sao Paulo, Brazil, concerning the relationship between urban environmental health and sustainability (McGranahan et al. 2001). In their study, narrow meanings of "environment", "sustainability" and "health" have been defined in largely physical terms. In the context of sustainable development, these words sometimes include wider meaning. Therefore, definition of the technical terms is important. Their analysis areas have been expanded to include "green" and "blown" concepts and necessity of a more politically and ethically explicit approach has been argued.

The Government of India has established ambitious goals to provide potable drinking water and to reduce infant mortality rate across the country in the conditions of lacking or completely missing infrastructure for providing safe drinking water and effective management of human wastes (Pattanayak et al. 2005). The World Bank has assisted Jalswarajya Project in Maharashtra State to meet these goals. The Project is supported by voluntary participation by communities. Water supply and sanitation services have been provided by project administrators.

A demand-driven institution, the Bolivian Social Investment Fund, has been established in 1991 after the Emergency Social Fund established in 1986 to improve coverage and quality of basic services in education, health, water and sanitation (Newman et al. 2002). A comparative evaluation of water and sanitation interventions have been conducted between two intervention areas, five provinces in Chaco and 17 provinces in Resto Rural, and control areas. Comparison has been conducted using a baseline dataset in 1993 and a follow-up data in 1997–1998. Distance to water sources and under-age-five mortality have been reduced significantly in the two intervention areas, and sanitation facilities improvement has been found in Resto Rural. Drinking water quality have not improved significantly.

Qualitative analyses on financial, institutional and managerial aspects of water and sanitation sectors in the world has been conducted (Ehrhardt et al. 2007). Organisational frameworks and regulations of water and sanitation sectors should be prepared to be appropriate for unique social and economic conditions by country. There have been examples of successes and failures among water and sanitation sector development in some countries and areas including Botswana, Burkina Faso, Colombia, the United Kingdom, Guyana, New Zealand, Australia, the United States of America, Senegal, Vanuatu, Bolivia, Phnom Penh, Manila, Florida, Buenos Aires, and municipalities in the Netherlands, New Zealand, and Scandinavia.

6.3 Resources and Profits of Sanitation Development

Recent studies have revealed that necessary resources for sanitation development are not only technologies and investments but also an institutional framework and appropriate regulations (Fig. 6.4).

World annual investments in sanitation sector for the "off-track countries" have been estimated as US$3.8 billion. Total of official development assistance (ODA) to the water and sanitation sectors by Development Assistance Committee countries and multilateral agencies has increased to more than US$6 billion in 2005 after declines in the 1990s (World Water Assessment Programme 2009). Total amount of the annual commitments of ODA from bilateral and multilateral agencies in water and sanitation sectors has been from US$3.1–4.4 billion in 2004–2006. The amount consists of 5.4–6.2 % of total ODA investments. In Uganda, water supply service coverage has increased from 47 % in 1998 to 71 % in 2007 using combination of several measures (World Water Assessment Programme 2009: Box14.23). The measures include timely and rational expansion of water production and network facilities, an optimal mixture of technology alternatives, introducing and subsidizing public access to water connections, innovative capital financing mechanisms, output-based investment approaches to strengthen service targeting, community-based approaches and mainstreaming consumer preferences, stakeholder coordination forums at different levels, and equitable providing water and sanitation infrastructures with an emphasis on sanitation. Income and profit of water delivery service after depreciation have increased from 21.9 to 70.4 billion Uganda Shillings and from −2.0 to 6.5 billion Uganda Shillings, respectively, in 1998–2007. Income and profit per employee have also increased.

Sanitation benefit in the world has been estimated as US$35 billion, about nine times benefit of the investments, i.e. US$9 benefits from US$1 investments (UN Water 2008). The sanitation benefits have comprised of 90.0 % time savings due to access to improve sanitation, 5.0 % deaths avoided, 3.1 % productive healthy days gained by people with avoided illness, 1.6 % health sector benefits due to avoided illness, and 0.2 % patients' expenses avoided.

6.4 Japanese ODA in Water and Sanitation Sectors

Japan has implemented US$4.6 billion of bilateral ODA in the fields of water and sanitation in 2000–2004, which corresponded to 41 % of total bilateral donors' ODA amounts in the fields (Ministry of Foreign Affairs 2006; Harada 2010). Besides large amount of Japanese financial contribution in water and sanitation sectors in the world, such contribution does not seem to be regarded as very important or beneficial in the water and sanitation sectors. Harada (2010) pointed out Japanese ODA in sanitation sector have concentrated to activated sludge technology which is popular in centralized WWTPs in Japan whereas "appropriate technology"[2] is demanded in developing countries, and emphasized importance of "self-sustainable"[3] aspect of "appropriate technology".

Follow-up of the intervention projects are also important. Follow-up survey of an integrated water supply, sanitation and hygiene (WSH) education intervention project has been conducted for the project operated by the International Centre for Diarrhoeal Disease Research, Bangladesh, in 1983–1987 (Hoque et al. 1996). The pumps in good functional condition have decreased with time, 94 % in 1987 and 82 % in 1992. Latrines functional have also decreased chronologically, 93 % in 1987 and 64 % in 1992. However, percentage of adults using these sanitary latrines in the projected areas has been larger, 84 % in 1992 comparing to 7 % in the control area without project. Infectious disease knowledge has been poor in both intervention and control areas, whereas diarrhoeal disease prevalence in control areas has been twice as large as that intervention areas. Although people's awareness on sanitation has continued to be high in intervention area, hardware cannot be properly maintained. Self-sustainability or additional supports should be considered after investment projects.

6.5 Japanese Sanitation Sector Development with a New Initiative for Asia and Pacific Regions

Japan Sanitation Consortium (JSC) has been established in October, 2009, which is aimed to work as a knowledge hub of sanitation sector in Japan (Japan

---

[2] The words are described in English in the preface article of Journal of Japan Society on Water Environment in Japanese.

[3] The words are also described in English in the preface article.

Sanitation Consortium 2009; Asia–Pacific Water Forum 2009). The concept of JSC has been approved at Asia Pacific Water Forum in June, 2009. JSC is consisted of Sewerage Business Management Centre (Sewerage Business Management Centre 2008), Japan Environmental Sanitation Centre, Japan Sewerage Works Association, Japan Education Centre for Environment Sanitation and so on and supported by Ministry of Land, Infrastructure, Transportation and Tourism (MLIT) and Ministry of the Environment (MoE). The purposes of JSC are to make a network of organizations in the field of sanitation, to accumulate and disseminate knowledge and information in the field of sanitation and to hold those in common. JSC is financially assisted by Asian Development Bank (ADB). JSC will prepare a networking of the countries in the region, hold international seminars, conduct research activities, and advise and assist activities of ADB and Japan International Cooperation Agency (JICA) in the field of sanitation.

For cost, construction cost of combined *johkasou* is larger than on-site sanitation in the other Asian countries (Tsuzuki 2010a, b). Construction cost reduction by domestic industries may be possible when recent development of low-price vehicles in a developing country is considered. For pollutant discharge reduction benefit among miscellaneous benefits, recent research and development of combined *johkasou* in Japan successfully achieved high removal efficiency of pollutants which can meet effluent concentration standards of 10 g-BOD m^{-3}, 10 g-TN m^{-3} and 1 g-TP m^{-3} (Kitao 2006). Japanese government and industry sectors have been tried to proliferate *johkasou* technology especially since the Third Water Forum held at Kyoto, Japan, in 2003 (Ministry of the Environments 2006; Katayama 2006). Some amounts of combined *johkasou* have been introduced to Eastern Europe, Indonesia, Malaysia, Thailand, China, Middle East and the United States of America. Japanese industries seem to leave on-site sanitation markets in China and Malaysia, however, domestic industries of these countries are dealing with *johkasou* type on-site treatment systems. There are about 500 private companies dealing with *johkasou* type on-site wastewater treatment systems in China. In Korea, official regulations and development systems of combined *johkasou* type on-site wastewater treatment systems have been developed, and there have been about 90,000 combined *johkasou* operated to treat both black and gray water.

In China, on-site wastewater treatment systems called *huafenchi* is installed in houses, which is mandated by the local government regulations regardless of existence of centralized WWTPs (Oh 2005). *Huafenchi* is often consisted of concrete structures and treats only black water whereas *johkasou* is made of plastic package and treats both black and gray water. *Huafenchi* is similar to septic tank in the world sanitation perspectives. There are sometimes clogging problems during *huafenchi* operation. Most Chinese people are not willing to pay for Japanese type *johkasou*. However, prediction of water resource shortage will increase possibility of technology transfer of Japanese combined *johkasou* because of effluent water quality which is suitable for reuse. Possible customers of combined *johkasou* will be collective housings, office buildings, hotels, recreational facilities and exclusive residential districts.

JSC activity will broaden activity of Japanese sanitation sector in Asia and Pacific countries especially in the fields of knowledge and information not only sewerage works and combined *johkasou* on-site treatment systems but also other sanitation alternatives.

6.6 Further Reading Materials on Low-Cost Sanitation

The followings are further reading materials on water and sanitation especially low-cost sanitation.

(1) Mara DD: Sanitation Challenge_paper ref. no. 5_Duncan Mara. http://eprints. whiterose.ac.uk/9218/1/SanitationChallenge.pdf

(2) Mara D, Alabaster G (2008) A new paradigm for low-cost urban water supplies and sanitation in developing countries. Water Policy http://www. hks.harvard.edu/var/ezp_site/storage/fckeditor/file/pdfs/centers-programs/ centers/cid/ssp/docs/events/workshops/2009/water/Mara_2008_Water_ Policy.pdf

(3) WHO (Hutton & Halle): Evaluation of the costs and benefits of water and sanitation improvements at the global level. http://www.who.int/water_ sanitation_health/wsh0404.pdf

(4) Haller L, Hutton G, Bartram J (2007) Estimating the costs and health benefits of water and sanitation improvements at global level. Journal of Water and Health. http://www.iwaponline.com/jwh/005/0467/0050467.pdf

(5) Water, sanitation and hygiene (WASH) Resources. http://washresources. wordpress.com/category/topics/sanitation/ecological-sanitation/

(6) Abeysuriya K: Cost Recovery for Urban Sanitation in Asian Countries. http:// www.anzsee.org/anzsee2005papers/Abeysuriya_Urban_sanitation_in_ Asian_Countries.pdf

(7) Paterson C, Mara D, Curtis T (2007) Pro-poor sanitation technologies. Geoforum. http://linkinghub.elsevier.com/retrieve/pii/S0016718506001333

(8) Agudelo C: Multi-criteria framework for the selection of urban sanitation systems. http://www.switchurbanwater.eu/outputs/pdfs/PAP_Multi-criteria_ framework_for_urban_sanitation_systems.pdf

(9) McKibbin JL: Valuing sustainable sanitation. http://www.isf.uts.edu.au/ publications/McKibben2008valuingsustainablesanitation.pdf

(10) Simplified sewerage (Wikipedia, the free encyclopedia). http://en.wikipedia. org/wiki/Simplified_sewerage

(11) Infrastructure Development Institute, Japan (2004) A draft Guideline for low-cost sewerage systems in developing countries (in Japanese and English). http://www.idi.or.jp/english/32ube.htm

(12) WHO (2010) UN-water global annual assessment of sanitation and drinking water (GLAAS) 2010. http://www.unwater.org/downloads/UN-Water_

GLAAS_2010_Report.pdf, http://www.unwater.org/activities_GLAAS2010.
html
(13) United Nations (2010) The millennium development goals Report 2010.
http://www.un.org/millenniumgoals/,   http://www.un.org/millenniumgoals/
environ.shtml

References

Asia-Pacific Water Forum (2009) Welcome to regional network for water knowledge hubs! http://
    www.apwf-knowledgehubs.net/
Bartram J, Lewis K, Lenton R, Wright A (2005) Focusing on improved water and sanitation for
    health. Lancet 365:810–812
Checkley W, Gilman RH, Black RE, Epstein LD, Cabrera L, Sterling CR, Moulton LH (2004)
    Effect of water and sanitation on childhood health in a poor Peruvian peri-urban community.
    Lancet 363:112–118
Ehrhardt D, Groom E, Halpern J, O'Connor S (2007) Economic regulation of urban water and
    sanitation services: some practical lessons. World Bank, Washington, DC, p 36. Water Sector
    Board Discussion Paper Series, Paper No. 9. http://siteresources.worldbank.org/INTWSS/
    ResourcesWS9_EconReg_C.pdf. Accessed 7 April 2010
Fewtrell L, Kaufmann RB, Kay D, Enanoria W, Haller L, Colford JM Jr (2005) Water, sanitation,
    and hygiene interventions to reduce diarrhoea in less developed countries: a systematic review
    and meta-analysis. Lancet Infect Dis 5(1):42–52
Franceys R, Pickford J, Reed R (1992) A guide to the development of on-site sanitation. WHO,
    Geneva, p 229. http://www.who.int/water_sanitation_health/hygiene/envsan/onsitesan/en/
Harada H (2010) Clear and present danger: collapse of water environment in developing regions,
    and its restoration by appropriate technology. Jpn Soc Water Environ 32(9):1. (in Japanese)
Hoque BA, Juncker T, Sack RB, Ali M, Aziz KM (1996) Sustainability of a water, sanitation and
    hygiene education project in rural Bangladesh: a 5-year follow-up. Bull World Health Organ
    74(4):431–437
Hutton G, Hallur L (2004) Evaluation of the costs and benefits of water and sanitation
    improvements at the global level, water, sanitation and health, protection of the human
    environment. WHO, Geneva, p 87
Infrastructure Development Institute, Japan (2004) Guidelines for low-cost sewerage systems in
    developing countries (draft), p 270. (in Japanese and in English)
Japan Sanitation Consortium (2009) Reference material: Japan Sanitation Consortium (JSC) will
    be established. http://www.mlit.go.jp/common/000050452.pdf. Accessed 9 April 2010. (in
    Japanese)
Katayama T (2006) *Johkasou* in the world. Presentation material for 19th Specialist Meeting on
    Johkasou, Wastes and Recycle Section, Central Environment Council, 22 Sept 2006. (in
    Japanese)
Kitao T (2006) Structure and maintenance of small johkasou. Japan Education Centre for
    Environment Sanitation, Tokyo, p 196. (in Japanese)
Lofrano G, Brown J (2010) Wastewater management through the ages: a history of mankind. Sci
    Total Environ. doi:10.1016/j.scitotenv.2010.07.062
Lyonnaise Des Eaux (1998). Alternative solutions for water supply and sanitation in areas with
    limited financial resources, prepared for UNEP
Mara D (1996a) Low-cost sewerage. Wiley, New York, p 238
Mara D (1996b) Low-cost urban sanitation. Wiley, New York, p 240

McGranahan G, Jacobi P, Songsore J, Surjadi C, Kjellen M (2001) The citizens at risk: from urban sanitation to sustainable cities (Risk, Society and Policy Series). Earthscan Publication Ltd, London, p 240

Ministry of Foreign Affairs, Japan (2006) Japan announces a new ODA Initiative: "Water and Sanitation Broad Partnership Initiative" (WASABI). http://mofa.go.jp/announce/announce/2006/3/0320.html

Ministry of the Environments, Japan (2006) Minutes of 19th Specialist Meeting on Johkasou, Wastes and Recycle Section, Central Environment Council, 22 Sept 2006. (in Japanese)

Newman J, Pradhan M, Rawlings L, Riddder G, Coa R, Evia JL (2002) An impact evaluation of education, health and water supply investments by the Bolivian Social Investment Fund. The World Bank Economic Review 16(2):241–274. Cited in Poulos et al. (2006)

Norman GM, Gravani RB (2006) Principles of food sanitation, 5th edn. Springer, New York, p 413

Oh H (2005) Research on johkasou technology transfer in China: case study of Harbin City. Master Thesis of Toyo University, Japan. (in Japanese)

Pattanayak SK, Yang JC, Patil S, Poulos C, Jones K, Kleinau E, Corey C, Kwok R (2005) Environmental Health Impacts of Water Supply, Sanitation and Hygiene Interventions in Rural Maharashtra, India. Study Protocol, Submitted to The World Bank. Cited in Poulos et al. (2006)

Poulos C, Pattanayak SK, Jones K (2006) A guide to water and sanitation sector impact evaluations. The World Bank, Washington, p 44. http://siteresources.worldbank.org/INTISPMA/Resources/383704-1146752240884/Doing_ie_series_04.pdf

Prüss A, Kay D, Fewtrell L, Bartram J (2002) Estimating the burden of disease from water, sanitation, and hygiene at a global level. Environ Health Perspect 110:537–542

Serageldin I (1994) Water supply, sanitation, and environmental sustainability: the financing challenge. The World Bank, Washington, p 42. http://www-wds.worldbank.org/external/default/WDSContentServer/WDSP/IB/1994/11/01/000009265_3970716143355/Rendered/PDF/multi_page.pdf

Sewerage Business Management Centre, Japan (2008) Sewerage business management centre homepage. http://www.sbmc.or.jp/english/index.html

Shifrin NS (2005) Pollution management in twentieth century. J Environ Eng ASCE 131:676–691

Tsuzuki Y (2010a) Chapter 5: Domestic wastewater pollutant discharge and pollutant load water quality in the ambient water in developed and developing countries. In: Ertuð K, Mirza I (eds) Water quality: physical, chemical and biological characteristics. Nova Science Publishers Inc, New York, pp 125–164 (277 pages), ISBN: 978-1-60741-633-3

Tsuzuki Y (2010b) Municipal wastewater pollutant discharge reduction with community participation in the Yamato-gawa River Basin, Japan, International Water Association (IWA) Specialist Group on Diffuse Pollution Newsletter No. 31. http://www.iwahq.org/Home/Networks/Specialist_groups/List_of_groups/Diffuse_Pollution Accessed Sept 2010

Tsuzuki, Y. (2011) Chapter 6: Sanitation Development and Roles of Japan, 179–202, in Joel M. McMann ed. Potable Water and Sanitation, Nova Science Publishers Inc, New York, p 266. ISBN: 978-1-61122-319-4

Tsuzuki Y, Koottatep T, Rahman MM, Ahmed F (2008a) Water quality in the ambient water and domestic wastewater pollutant discharges in the developing countries: possibility of combined johkasou exports, Joukasou Kenkyu. J Domest Wastewater Treat Res 20(1):1–13. (in Japanese with English abstract)

Tsuzuki Y, Koottatep T, Ahmed F, Rahman MM (2008b) Domestic wastewater pollutant discharge and pollutant load in the tidal area of the ambient water in developing countries: survey results in autumn and winter in 2006. J Global Environ Eng 13:121–133, Japan Society of Civil Engineers. http://www.wsscc.org/resources/resource-publications/water-quality-and-pollutant-load-ambient-water-and-domestic

UNICEF and WHO (2012) Progress on drinking water and sanitation 2012, p 59

UN Water (2008) Fact sheet 2: sanitation generates economic benefit. http://esa.un.org/iys/docs/
    IYS%20Advocacy%20kit%20ENGLISH/Fact%20sheet%202%20pdf.   Accessed   10   April
    2010
Water Supply and Sanitation Collaboration Council and World Health Organisation (2005)
    Sanitation and hygiene promotion: programming guidance, p 84
WHO/UNICEF Joint Monitoring Programme for Water Supply and Sanitation (2000) Global
    water supply and sanitation assessment 2000 Report, p 80
WHO/UNICEF Joint Monitoring Programme on Water Supply and Sanitation (2010) Progress on
    sanitation and drinking-water: 2010 Update, p 60. http://www.who.int/water_sanitation_
    health/publications/9789241563956/en/index.html
World Health Organisation (2004) Water, sanitation and hygiene links to health: Facts and figures
    updated.   http://www.who.it/water_sanitation_health/publications/facts2004/en/index.html.
    Accessed Nov 2004
World Water Assessment Programme (2009) The united nations world water development report
    3: water in a changing world. UNESCO and Earthscan, p 349. http://www.unesco.org/water/
    wwdr/wwdr3/side_publications.shtml

Index